A-Level
Mathematics

Exam Board: OCR

A-Level Maths exams are a piece of cake... *rock* cake, that is. [*Silence.*] They're hard, is what we're trying to say. So you're going to need plenty of practice to do well.

By incredible coincidence, this CGP book is bursting with exam-style questions for every A-Level topic — there's even a full set of practice papers at the end to make sure you're ready for the real exams. Sounds like just your sort of thing, actually.

What else can we tell you? Well, everything's been crafted to cover both years of the latest OCR course, and we've included step-by-step solutions and full mark schemes at the back. Go on, get your teeth into it.

A-Level revision? It has to be CGP!

Published by CGP

Editors:
Chris Corrall, Sammy El-Bahrawy, Will Garrison, Simon Little, Alison Palin, Andy Park, David Ryan, Caley Simpson

Contributors:
Kevin Bennett, Mark Moody, Kieran Wardell, Charlotte Young

ISBN: 978 1 78294 742 4

With thanks to Shaun Harrogate and Dawn Wright for the proofreading.
With thanks to Emily Smith for the copyright research.

Clipart from Corel®
Printed by Elanders Ltd, Newcastle upon Tyne

Based on the classic CGP style created by Richard Parsons.

Contents

Use the tick boxes to check off the topics you've completed.

Exam Advice

Good exam technique can make a big difference to your mark, so make sure you read this stuff carefully.

Get familiar with the **Exam Structure**

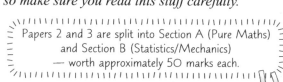

Papers 2 and 3 are split into Section A (Pure Maths) and Section B (Statistics/Mechanics) — worth approximately 50 marks each.

For **A-level Mathematics**, you'll be sitting **three papers**.

Pure Mathematics (01) 2 hours 100 marks	**33.33%** of your A-level	Covers the topics tested in **Section One** of this book.
Pure Mathematics and Statistics (02) 2 hours 100 marks	**33.33%** of your A-level	Covers the topics tested in **Section One & Two** of this book.
Pure Mathematics and Mechanics (03) 2 hours 100 marks	**33.33%** of your A-level	Covers the topics tested in **Sections One & Three** of this book.

Some formulas are given in the *Question Paper*

The exam paper will contain a **formula sheet** in the front that lists some of the formulas you might need. The formulas you'll be given are shown on pages 164-166 of this book.
You don't need to learn these formulas but you do need to know **how to use** them.

Manage Your Time sensibly

1) The **number of marks** tells you roughly **how long** to spend on a question — you've got just over a minute per mark in the exam. If you get stuck on a question for too long, it may be best to **move on** so you don't run out of time for the others.

2) You don't have to work through the paper **in order** — you could leave questions on topics you find harder until last.

Get *Familiar* with the *Large Data Set*

Throughout your course you'll be working with the **large data set**. This is a group of **tables** containing information about **how people travel to work** (e.g. car, bus, train, etc.), and a breakdown of the population by **age**. The large data set will only be used in Paper 2 for A-level Maths.

Questions in this paper might:

- Assume that you're familiar with the **terminology** and **contexts** of the data.
- Use **summary statistics** based on the large data set — this might reduce the time needed for some calculations.
- Include **statistical diagrams** based on the large data set.
- Be based on a **sample** from the large data set.

You might be expected to know specific details about the large data set — such as differences between the Local Authorities.

Watch out for *Modelling* and *Problem-Solving* questions

The A-level Maths course has a few **overarching themes** — **proof, problem-solving** and **modelling**. The first topic in this book covers proof (and there are other proof questions dotted throughout the book). Problem-solving and modelling questions are covered throughout the book, but they can be trickier to spot.

- **Problem-solving** questions involve skills such as **combining** different areas of maths or **interpreting information** given to identify what's being asked for. They're often worth a **lot of marks**, as there's a lot of maths involved in them.

- **Modelling** questions involve using maths to represent **real-life situations**. You might be asked to think about the **validity** of the model (how realistic it is) or to interpret values **in context**.

Proof

Welcome one and all to this wondrous (if I do say so myself) A-Level practice book. First up is a real tough cookie — proof. But, if you prove yourself worthy on this section, you'll be well on your way to success...

1   Prove the following statement:

For all integers n, $n^3 - n$ is always even.

(2 marks)

2   Prove that there is no largest integer.

(3 marks)

3   Prove that the product of any two distinct prime numbers has exactly four factors.

(3 marks)

4   Prove that if a and b are rational numbers, the number $c = a - b$ is also rational.

(3 marks)

5   Prove that $n^3 + 2n^2 + 12n$ always has a factor of 8 when n is even.

(3 marks)

Proof

6 A student makes the following statement:

"If x^3 is even, then x is even."

She attempts to prove the claim as follows:

"Let $x = 2k$. Then $x^3 = (2k)^3 = 8k^3 = 2(4k^3)$ which must be even, therefore the claim is true."

a) Explain why the student's proof is not valid.

....

....

....

(1 mark)

b) Prove the student's statement.

(4 marks)

7 Riyad claims that, "if x and y are both irrational, then $\dfrac{x}{y}$ is also irrational".

a) Disprove Riyad's claim with a counter-example.

(1 mark)

Riyad goes on to claim that "any non-zero rational number multiplied by any irrational number is irrational."

b) Prove Riyad's claim by contradiction.

(3 marks)

When you're doing a proof in an exam, it's really important that it's laid out in a clear and logical manner. If it's a proof by contradiction, state the assumption you've made and then show how this assumption leads to a contradiction. Once you have the contradiction, don't just stop — finish with a concluding line starting "hence" or "therefore".

Score

23

Algebra and Functions 1

Algebra is a pretty important part of maths — so it's a good idea to get to grips with it now. First up, Algebra and Functions 1, where you can practise all things surd-like and fraction-y. The excitement is almost palpable.

1 Simplify $\dfrac{a^6 \times a^3}{\sqrt{a^4}} \div a^{\frac{1}{2}}$. Give your answer in the form $a^{\frac{p}{q}}$.

....

(2 marks)

2 Find the value of x such that:

a) $27^x = 3$

$x =$ ....
(1 mark)

b) $27^x = 81$

$x =$ ....
(1 mark)

3 Fully simplify $\dfrac{(3ab^3)^2 \times 2a^6}{6a^4 b}$.

....

(2 marks)

4 Show that $\dfrac{(5 + 4\sqrt{x})^2}{2x}$ can be written as $\dfrac{25}{2}x^{-1} + Px^{-\frac{1}{2}} + Q$, where P and Q are integers.

(3 marks)

Section One — Pure Maths

Algebra and Functions 1

5   Express $\left(5\sqrt{5} + 2\sqrt{3}\right)^2$ in the form $a + b\sqrt{c}$, where a, b and c are integers to be found.

....

(4 marks)

6   Rationalise the denominator of $\dfrac{10}{\sqrt{5} + 1}$. Give your answer in its simplest form.

....

(3 marks)

7   Express $\dfrac{4 + \sqrt{2}}{2 + \sqrt{2}}$ in the form $a + b\sqrt{2}$, where a and b are integers.

....

(3 marks)

8   Express $\dfrac{(x^2 - 9)(3x^2 - 10x - 8)}{(6x + 4)(x^2 - 7x + 12)}$ as a fraction in its simplest form.

....

(3 marks)

Algebra and Functions 1

9 For this question, give all answers as fractions in their simplest form.

a) Simplify $\dfrac{x^2 + 5x - 14}{2x^2 - 4x}$.

....
(2 marks)

b) Using your answer to part a) or otherwise, write $\dfrac{x^2 + 5x - 14}{2x^2 - 4x} + \dfrac{14}{x(x - 4)}$ as a single fraction.

....
(3 marks)

10 Express $\dfrac{1}{x(2x - 3)}$ in partial fractions.

....
(3 marks)

11 Express $\dfrac{6x - 1}{x^2 + 4x + 4}$ in partial fractions.

⌐‾‾‾‾‾‾‾‾‾‾‾‾‾‾‾‾‾‾‾‾‾‾‾‾‾‾‾‾‾¬
  You'll have to factorise the denominator first.
∟_____┘

....
(4 marks)

EXAM TIP

In an exam, you might not always be asked to write an expression as partial fractions, but if the denominator is a product of two linear factors, it's usually a good place to start. There's so much algebra to cover in A-Level Maths, I've decided to split it up into a few parts. This is the end of Part 1. Next, with crushing inevitability, comes Part 2...

Score

34

Algebra and Functions 2

Algebra and Functions Part 2 will take you on a whistle-stop tour of quadratic equations, make a slight detour into the world of simultaneous equations and inequalities, and finish with everyone's favourite — cubics.

1   Given that the equation $3jx - jx^2 + 1 = 0$, where j is a constant, has no real roots, find the range of possible values of j.

....

(3 marks)

2 $f(x) = \dfrac{1}{x^2 - 7x + 17}$

    a)   Express $x^2 - 7x + 17$ in the form $(x - m)^2 + n$, where m and n are constants.

....

(3 marks)

    b)   Hence find the maximum value of $f(x)$.

....

(2 marks)

3   Find the possible values of k if the equation $g(x) = 0$ is to have two distinct real roots, where $g(x)$ is given by $g(x) = 3kx^2 + kx + 2$.

....

(3 marks)

4   Solve the equation $x^6 = 7x^3 + 8$.

....

(4 marks)

Algebra and Functions 2

5 A scientist working at a remote Arctic research station monitors the temperature during the hours of daylight.
For a day with 9 hours of sunlight, she models the temperature using the function $T = 10h - h^2 - 27$,
where T is the temperature in °C and h is the time in hours since sunrise.

a) **(i)** Express $T = 10h - h^2 - 27$ in the form $T = -(m - h)^2 + n$, where m and n are integers.

....

(3 marks)

(ii) Hence show that T is always negative according to this model.

(1 mark)

b) **(i)** State the maximum temperature predicted by this model,
and state the number of hours after sunrise at which it will occur.

Temperature = ... °C, time = ... hours after sunrise

(2 marks)

(ii) Sketch the graph of $T = 10h - h^2 - 27$ on the axes below.
Mark clearly on your graph the temperature at sunrise.

(2 marks)

6 Solve the simultaneous equations $y + x = 7$ and $y = x^2 + 3x - 5$.

$x = $ .... $y = $ .... or $x = $ .... $y = $ ....

(4 marks)

Section One — Pure Maths

Algebra and Functions 2

7   The curve C has equation $y = -x^2 + 3$ and the line l has equation $y = -2x + 4$.

   a)   Find the coordinates of the point (or points) of intersection of C and l.

   ....
   (3 marks)

   b)   Sketch the graphs of C and l on the axes below, clearly showing
          where the graphs intersect the x- and y- axes.

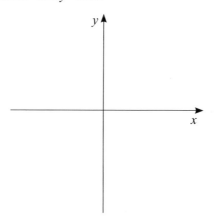

   (5 marks)

8   Draw and label the region that satisfies the inequalities $y \geq x + 2$ and $4 - x^2 > y$.

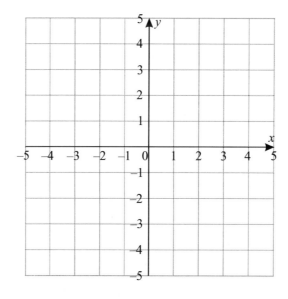

   (3 marks)

9   Solve the inequality $x^2 - 8x + 15 > 0$. Give your answer in set notation.

   ....
   (4 marks)

Algebra and Functions 2

10 $(x-1)(x^2 + x + 1) = 2x^2 - 17$

a) Rewrite the equation above in the form $f(x) = 0$, where $f(x)$ is of the form $f(x) = ax^3 + bx^2 + cx + d$.

....

(2 marks)

b) Show that $(x + 2)$ is a factor of $f(x)$.

(2 marks)

c) Hence write $f(x)$ as the product of a linear factor and a quadratic factor.

....

(2 marks)

d) By completing the square, or otherwise, show that $f(x) = 0$ has only one root.

(2 marks)

11 A function is defined by $f(x) = 2x^3 - 7x^2 - ax + 10$. $(2x - 1)$ is a factor of $f(x)$.
Find the value of a and hence or otherwise solve the equation $2x^3 - 7x^2 - ax + 10 = 0$.

Use the factor theorem
to find the value of a.

$a = $ .... $x = $ ....

(6 marks)

EXAM TIP

Quadratics have a habit of popping up in exam questions where you least expect them (like in exponentials, trig equations or mechanics, not to mention in simultaneous equations, inequalities and cubics) — so make sure you can handle them. It's worth practising your factorising skills, as it could save you a lot of time in the exam, and time is ~~money~~ marks.

Score

56

Algebra and Functions 3

This section should appeal to your artistic side — you get to draw some beautiful graphs. Although don't panic if you're not artistic — there's some equally beautiful algebra coming up later in the form of the binomial expansion.

1 $f(x) = |2x + 3|$ and $g(x) = |5x - 4|$.

    a)   On the same axes, draw the graphs of $y = f(x)$ and $y = g(x)$,
        showing clearly where each graph meets the coordinate axes.

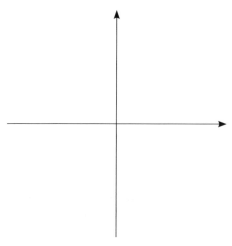

(2 marks)

    b)   Hence or otherwise solve the equation $f(x) = g(x)$.

....

(4 marks)

2 $f(x) = |4x + 5|$.

    a)   Find the possible values of $f(x)$ if $|x| = 2$.

....

(3 marks)

    b)   Find the values of x for which $f(x) \leq 2 - x$.

....

(3 marks)

    c)   Find the possible values of a constant A for which the equation $f(x) + 2 = A$ has two distinct roots.

....

(2 marks)

Algebra and Functions 3

3   Sketch the curve of $y = (x - 2)^2(x + 3)^2$ on the axes provided.
Show clearly any points of intersection with the x- and y-axes.

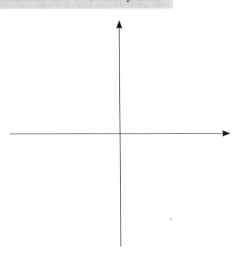

(3 marks)

4   The graph of $y = x^3 - 2x^2 + px$, for some constant p, crosses
the x-axis at the points $(1 - \sqrt{3}, 0)$, $(0, 0)$ and $(1 + \sqrt{3}, 0)$.

   a)   Sketch the graph of $y = x^3 - 2x^2 + px$, showing clearly any points of intersection with the axes.

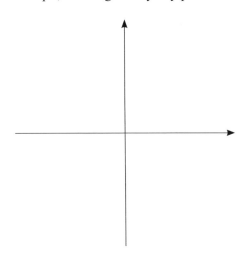

(2 marks)

   b)   Find the value of p.

            It'll help to factorise the
            original function first.

....

(2 marks)

5   Describe what happens to the curve $y = x^3$ to transform it into the curve $y = 2(x - 1)^3 + 4$.

....

....

....

(3 marks)

Algebra and Functions 3

6 A diver's position is modelled by the function $V = 2t^3 - 10t^2 + 8.5t + 7, \ 0 \leq t \leq 3.5$.
V is the vertical height of the diver's head above the surface of the pool, in metres.
t is the time in seconds from him starting his dive.
At the deepest point of the dive, the diver is 3.70 m (3 s.f.) below the surface of the pool.

a) Show that V can be written as $V = (2t + 1)(t - 2)(t - 3.5)$.

(2 marks)

b) Hence sketch the graph of $y = V(t)$ for $0 \leq t \leq 3.5$.
Label any points that the graph intercepts the axes.

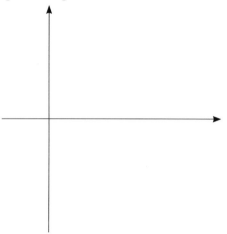

(3 marks)

c) How many seconds after starting the dive does the diver enter the pool?

.... s
(1 mark)

d) Given that the diver is 1.75 m tall, find the height of the diving board above the surface of the pool.

.... m
(2 marks)

e) The same diver then dives from a higher diving board. Darren suggests adapting the model for this diving board. The adapted model is $V = 2t^3 - 10t^2 + 8.5t + 10$.
Comment on the validity of this adapted model, giving reasons for your comments.

> Think about how the model has been adapted in terms of its graph.

....

....

....

(2 marks)

Algebra and Functions 3

7   Figure 1 shows a sketch of the function $y = f(x)$.
The function crosses the x-axis at $(-1, 0)$, $(1, 0)$ and $(2, 0)$,
and crosses the y-axis at $(0, 2)$.

Figure 1

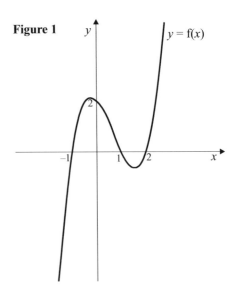

Sketch each transformation on the axes given below.
On each diagram, label any known points of intersection with the x- or y-axes.

a) $y = f\left(\frac{1}{2}x\right)$

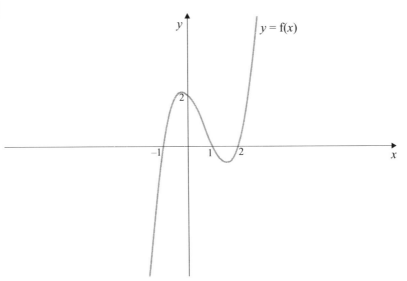

(3 marks)

b) $y = 2f(x - 4)$

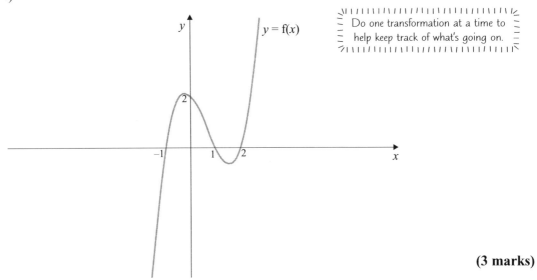

Do one transformation at a time to
help keep track of what's going on.

(3 marks)

Algebra and Functions 3

8 The graph on the right shows the function $y = f(x)$, $x \in \mathbb{R}$, with turning points $A(-1, -2)$ and $B(3, 2)$.

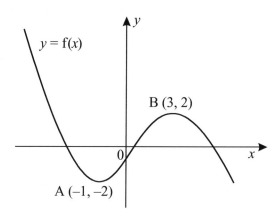

On separate axes, sketch the graphs of the following, clearly showing the coordinates of A and B where possible.

a) $y = f(-x) - 1$.

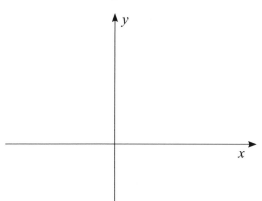

(3 marks)

b) $y = 3f(x + 2)$.

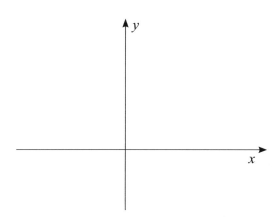

(3 marks)

9 The graph of the function $f(x)$ passes through points P(1, 2) and Q(3, 16). $f(x)$ is first translated 3 units up then 2 units right and finally reflected in the y-axis to form a new graph, $g(x)$.

a) Write down the equation of $g(x)$ in terms of $f(x)$.

....

(3 marks)

b) Write down the new coordinates of the points P and Q.

....

(2 marks)

Algebra and Functions 3

10 The functions f and g are defined as follows: $f(x) = 2^x, x \in \mathbb{R}$ and $g(x) = \sqrt{3x + 1}, x \geq -\frac{1}{3}$.

   a)   **(i)**   Find $gf(x)$.

....

(1 mark)

        (ii)   Hence solve $gf(x) = 5$.

....

(2 marks)

   b)   Find $g^{-1}(x)$ and state its domain and range.

....

(3 marks)

11 The functions f and g are defined as follows: $f(x) = \frac{1}{x^2}, x \in \mathbb{R}, x \neq 0$ and $g(x) = x^2 - k, x \in \mathbb{R}$, where k is a positive integer.

   a)   State the range of g. Give your answer in terms of k.

....

(1 mark)

   b)   Neither f nor g have an inverse. Explain why.

....

(1 mark)

   c)   **(i)**   Given that $gf(1) = -8$, find the value of k and hence find $fg(x)$, and write down the domain of the composite function fg.

....

(5 marks)

        (ii)   Hence solve $fg(x) = \frac{1}{256}$.

....

(4 marks)

EXAM TIP   Don't worry, graph sketches don't have to be perfect — as long as they're generally the correct shape, and any turning points and intersections are in the right places (and labelled if the question asks for it), you should get all the marks. It's always a good idea to sketch graphs in pencil in case you make a mistake (and make sure you have a rubber to hand just in case).

Score

68

Coordinate Geometry

Up, down, up, triangle, circle, circle, down, triangle... I'm afraid when it comes to coordinate geometry and circle equations, there are no cheat codes. So pens at the ready — you've got to do this the hard way.

1 The point A lies at the intersection of the lines l_1 and l_2, where the equation of l_1 is $x - y + 1 = 0$ and the equation of l_2 is $2x + y - 8 = 0$.

    a) Find the coordinates of point A.

....
(3 marks)

    b) The points B and C have coordinates $(6, -4)$ and $\left(-\frac{4}{3}, -\frac{1}{3}\right)$ respectively, and D is the midpoint of AC. Find the equation of the line through B and D in the form $ax + by + c = 0$, where a, b and c are integers.

....
(5 marks)

    c) Show that the triangle ABD is a right-angled triangle.

(3 marks)

2 The diagram shows a square ABCD, where point B has coordinates $(3, k)$. The line through points B and C has equation $-3x + 5y = 16$.

    a) Show that the line with equation $5x + 3y - 6 = 0$ is parallel to the line through points A and B.

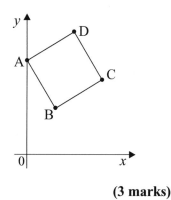

(3 marks)

    b) Find the area of square ABCD.

....
(5 marks)

Coordinate Geometry

3 The points A(2, 1) and B(0, –5) lie on a circle, where the line AB is a diameter of the circle.

a) Find the centre and radius of the circle.

centre = ..., radius = ...

(3 marks)

b) Show that the point (4, –1) also lies on the circle.

(2 marks)

c) Show that the equation of the circle can be written in the form $x^2 + y^2 - 2x + 4y - 5 = 0$.

(2 marks)

d) Find the equation of the tangent to the circle at point A, giving your answer in the form $y = mx + c$.

....

(3 marks)

4 The diagram shows a circle with centre P.
The line AB is a chord with midpoint M.

a) Show that $p = 5$.

The diagram shows a chord,
so think about which circle
property might apply.

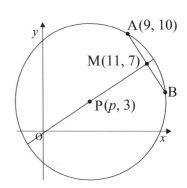

(5 marks)

b) Find the equation of the circle.

....

(3 marks)

Coordinate Geometry

5   The circle with equation $x^2 - 6x + y^2 - 4y = 0$ crosses the y-axis at the origin and the point A.

a)   Find the coordinates of point A.

....
(2 marks)

b)   Write the equation of the circle in the form $(x - a)^2 + (y - b)^2 = c$.

....
(3 marks)

c)   Write down the radius and the coordinates of the centre of the circle.

radius = ..., centre = ...
(2 marks)

d)   The tangent to the circle at point A meets the x-axis at point B. Find the exact distance AB.

....
(6 marks)

6 PQR is a triangle whose vertices all lie on the circumference of circle C, where $P = (3, 1)$, $Q = (0, 2)$ and $R = (1, 5)$. Find the equation of C. Give your answer in the form $(x - a)^2 + (y - b)^2 = c$.

Hint: the perpendicular bisectors of the sides of the triangle will meet at the centre of the circle.

....
(7 marks)

Coordinate Geometry

7 The curve below has parametric equations $x = 3\sin\theta$, $y = 4\cos\theta$.

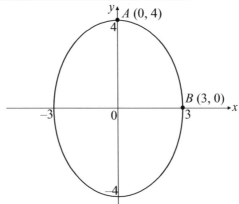

a) For $0 \le \theta \le \dfrac{\pi}{2}$, find the values of θ that correspond to the points A and B.

....

(2 marks)

b) Show that $y^2 = (4 + \dfrac{4x}{3})(4 - \dfrac{4x}{3})$.

> You'll need to use a trig
> identity for this one...

(3 marks)

8 Part of the path of a boat sailing around an island is modelled on a shipping chart with the parametric equations $x = t^2 - 7t + 12$, $y = t - 1$, where t is time in hours.

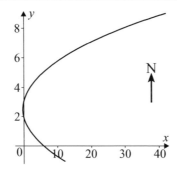

In this model, the western tip of the island has coordinates $(12, 2)$.
Calculate the length of time that the boat is further west than the tip of the island.

.... hours

(3 marks)

Coordinate Geometry

9 The curve C is defined by the parametric equations:

$$x = 1 - \tan\theta, \quad y = \tfrac{1}{2}\sin 2\theta, \quad -\tfrac{\pi}{2} < \theta < \tfrac{\pi}{2}.$$

a) P is the point on curve C where $\theta = \tfrac{\pi}{3}$. Find the exact coordinates of P.

....

(2 marks)

b) Point Q on curve C has coordinates $(2, -\tfrac{1}{2})$. Find the value of θ at Q.

....

(2 marks)

c) Using the identity $\sin 2\theta \equiv \dfrac{2\tan\theta}{1 + \tan^2\theta}$, show that the Cartesian equation of C is $y = \dfrac{1 - x}{x^2 - 2x + 2}$.

....

(3 marks)

10 A curve C has parametric equations $y = t^3 + t,\ x = 4t - 2$.

a) Find the coordinates of the point where the curve C crosses the y-axis.

....

(2 marks)

b) Find the coordinates of the points where the curve C intersects the line $y = \tfrac{1}{2}x + 1$.

....

(4 marks)

c) Find the Cartesian equation of the curve C in the form $y = ax^3 + bx^2 + cx + d$, where a, b, c and d are fractions.

$y = $ ....

(3 marks)

 EXAM TIP Working with parametric equations can be a real chore. When converting to Cartesian form, it's not always obvious how to rearrange the parametric equations, so you might have to try a couple of different ways — it's best to start with the equation that gives the simplest expression for t. Also, make sure you're up to snuff with your trig identities, as they'll come in handy.

Score

81

Sequences and Series 1

Time to test your knowledge of arithmetic and geometric sequences. Get ready for common differences, common ratios, recurrence relations, sums to infinity, sums to not-quite-infinity, sigma notation... I know you're gonna love it.

1 An arithmetic series has first term a and common difference d.
The value of the 12th term is 79, and the value of the 16th term is 103.

a) Find the values of a and d.

$a =$ .... $d =$ ....

(3 marks)

S_n is the sum of the first n terms of the series.

b) Find the value of S_{15}.

....

(2 marks)

The formula for S_n is given on the formula sheet (see p.164-166).

2 A sequence is defined by the recurrence relation $x_{n+1} = 3x_n - 4$, with first term $x_1 = k$.

a) Write a simplified expression for x_4 in terms of k.

....

(2 marks)

b) Determine whether the sequence is increasing or decreasing when $k = 1$.

(2 marks)

3 A sequence is defined by the recurrence relation: $h_{n+1} = 2h_n + 2$ when $n \geq 1$.

a) Given that $h_1 = 5$, find the values of h_2, h_3, and h_4.

....

(2 marks)

b) Calculate the value of $\sum_{r=3}^{6} h_r$.

....

(3 marks)

Sequences and Series 1

4 A geometric series $u_1 + u_2 + ... + u_n$ has 3rd term $\frac{5}{2}$ and 6th term $\frac{5}{16}$.

a) Find the formula for the nth term of the series.

....
(4 marks)

b) Find $\sum_{i=1}^{10} u_i$. Give your answer as a fraction in its simplest terms.

....
(2 marks)

c) Show that the sum to infinity of the series is 20.

(2 marks)

5 A geometric series has the first term 5 and is defined by: $u_{n+1} = 5 \times 1.7^n$.

a) Can the sum to infinity of this sequence be found? Explain your answer.

....
(1 mark)

b) Find the value of the 3rd and 8th terms.

$u_3 = $ .... $u_8 = $ ....
(2 marks)

6 In a geometric series with n^{th} term u_n, $a = 20$ and $r = \frac{3}{4}$.

Find values for the following, giving your answers to 3 significant figures where necessary:

a) S_∞

....
(2 marks)

b) u_{15}

....
(2 marks)

c) the smallest value of n for which $S_n > 79.76$.

Use the general formula for S_n, then start rearranging. You'll need to take logs to get n on its own.

....
(5 marks)

Sequences and Series 1

7 To raise money for charity, Alex, Chris and Heather were sponsored £1 for each kilometre they ran over a 10-day period. They receive sponsorship proportionally for partial kilometres completed.

Alex ran 3 km every day.

Chris ran 2 km on day 1 and on each subsequent day ran 20% further than the day before.

Heather ran 1 km on day 1, and on each subsequent day ran 50% further than the previous day.

a) How far did Heather run on day 5, to the nearest 10 metres?

.... km

(2 marks)

b) Show that day 10 is the first day that Chris runs further than 10 km.

(3 marks)

c) Find the total amount raised by the end of the 10 days, to the nearest penny.

....

(4 marks)

8 $a + ar + ar^2 + ar^3 + \ldots$ is a geometric series.
The second term of the series is −2 and the sum to infinity of the series is −9.

a) Show that $9r^2 - 9r + 2 = 0$.

(3 marks)

b) Find the possible values of r.

....

(2 marks)

c) Hence find the possible values of a.

....

(2 marks)

This is a top tip straight from the mouth of the examiners themselves, so you'd best listen up. There are a few different formulae to contend with here, so make sure you don't get flustered in the exam and get them muddled up. Don't get your u_n and your S_n mixed up or accidentally find the sum to infinity when you don't mean to — you're just throwing away marks that way.

Score

50

Sequences and Series 2

OK, now it's time for something a little different. The next few pages cover variations of binomial expansions — including cases where the power is a positive integer, a negative integer or even a fraction.

1 Use the binomial expansion formula to answer the following questions.

    a) Write down the first four terms in the expansion of $(1 + ax)^{10}$, $a > 0$.

....
(2 marks)

    b) Find the coefficient of x^2 in the expansion of $(2 + 3x)^5$.

....
(3 marks)

    c) Given that the coefficients of x^2 in both expansions are equal, find the value of a.

$a =$ ....
(1 mark)

2 The binomial expansion of $(j + kx)^6$ is $j^6 + ax + bx^2 + cx^3 + ...$

    a) Given that $c = 20\,000$, show that $jk = 10$ (where both j and k are positive integers).

(3 marks)

    b) Given that $a = 37\,500$, find the values of j and k.

$j =$ ....

$k =$ ....
(4 marks)

    c) Find b.

$b =$ ....
(2 marks)

Sequences and Series 2

3 $(1-x)^{-\frac{1}{2}} \approx 1 + Ax + Bx^2 + Cx^3$, where A, B and C are rational constants.

a) Find the values of the constants A, B and C.

$A = $ .... $B = $ .... $C = $ ....

(2 marks)

b) **(i)** Hence show that $(25 - 4x)^{-\frac{1}{2}} \approx \frac{1}{5} + \frac{2}{125}x + \frac{6}{3125}x^2 + \frac{4}{15\,625}x^3$ for small values of x.

(4 marks)

(ii) State the range of values of x for which the expansion from part (i) is valid.

....

(1 mark)

c) **(i)** Use your expansion from b) with a suitable value of x to show that $\frac{1}{\sqrt{20}} \approx \frac{447}{2000}$.

(3 marks)

(ii) What is the percentage error in this estimate? Give your answer to 3 significant figures.

.... %

(2 marks)

Sequences and Series 2

4 $f(x) = (27 + 4x)^{\frac{1}{3}}$, for $|x| < \dfrac{27}{4}$

    a)   Find the binomial expansion of f(x) up to and including the term in x^2.

....

(4 marks)

    b)   Hence find an approximation to $\sqrt[3]{26.2}$. Give your answer to 6 decimal places.

....

(2 marks)

5 $f(x) = \sqrt{\dfrac{1 + 3x}{1 - 5x}}$

    a)   **(i)**   Use the binomial expansion in increasing powers of x to show that $f(x) \approx 1 + 4x + 12x^2$.

(5 marks)

         (ii)   For what values of x is your expansion valid?

....

(2 marks)

    b)   Using the above expansion with $x = \dfrac{1}{15}$, show that $\sqrt{1.8} \approx \dfrac{33}{25}$.

(2 marks)

Sequences and Series 2

6 $f(x) = \dfrac{2 - 18x}{(5 + 4x)(1 - 2x)^2}$

a)   Given that $f(x)$ can be expressed in the form $f(x) = \dfrac{A}{(5 + 4x)} + \dfrac{B}{(1 - 2x)} + \dfrac{C}{(1 - 2x)^2}$,
find the values of A, B and C.

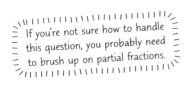

If you're not sure how to handle this question, you probably need to brush up on partial fractions.

$A =$ .... $B =$ .... $C =$ ....

(4 marks)

b)   Hence find the binomial expansion of $f(x)$, up to and including the term in x^2.

....

(7 marks)

c)   Claire states that the expansion above is valid for $\dfrac{1}{2} < x < \dfrac{5}{4}$.
Explain the error that Claire has made and state the correct range.

....

....

....

(2 marks)

EXAM TIP   As well as being happy with doing the expansions, make sure you're confident with the other things examiners often ask — such as approximations and ranges of validity. When you have several expansions you should calculate the valid ranges of each expansion, then the overall expansion will be valid for the narrowest of the two (or more) ranges.

Score

55

Trigonometry

You'd think that there's only so much you can do with a three-sided shape, right? Well, think again...
It's time to delve deeper into the twisted triangular world of trig, and discover new ways of measuring angles.

1 The diagram shows a sector of a circle of radius r cm and angle 120°.
The length of the arc of the sector is 40 cm.

Find the area of the sector to the nearest square centimetre.

.... cm²

(5 marks)

2 A new symmetrical mini-stage is to be built according to the design below.
The design consists of a rectangle of length q metres and width $2r$ metres,
two sectors of radius r and angle θ radians (shaded), and an isosceles triangle.

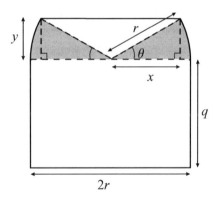

a) **(i)** Show that distance x is given by $x = r\cos\theta$.

(1 mark)

(ii) Find a similar expression for distance y.

....

(1 mark)

b) Find, in terms of r, q and θ, expressions for the perimeter P, and the area A, of the stage.

....

(4 marks)

c) Given that the perimeter of the stage is to be 40 m and $\theta = \dfrac{\pi}{3}$,

show that A is given by $A = 40r - kr^2$, where $k = 3 - \dfrac{\sqrt{3}}{4} + \dfrac{\pi}{3}$.

(4 marks)

Trigonometry

3   Sketch the graphs of $y = \sin x$ and $y = \sin \frac{x}{2}$ in the range $0 \leq x \leq 4\pi$ on the same set of axes, showing the points at which the graphs cross the x-axis.

(3 marks)

4   A sheep pen is modelled as a triangle with side lengths of 50 m, 70 m and 90 m.

     a)   Find the area of the sheep pen to the nearest square metre.

.... m²

(5 marks)

     b)   Comment on the accuracy of the model.

....

....

(1 mark)

5   Solutions to this question based entirely on graphical or numerical methods are not acceptable.

     Find all the values of x, in the interval $0° \leq x \leq 180°$, for which $7 - 3\cos x = 9\sin^2 x$.

....

(5 marks)

Trigonometry

6 Adam and Bethan have each attempted to solve the equation $\sin 2t = \sqrt{2} \cos 2t$ for the range $-90° < t < 90°$. Their working is shown below.

Adam

$\sin 2t = \sqrt{2} \cos 2t$

$\tan 2t = \sqrt{2}$

$\tan t = \dfrac{\sqrt{2}}{2}$

$t = 35.26...°$

Bethan

$\sin 2t = \sqrt{2} \cos 2t$

$\sin^2 2t = 2 \cos^2 2t$

$1 - \cos^2 2t = 2 \cos^2 2t$

$\cos^2 2t = \dfrac{1}{3}$

$\cos 2t = \pm \dfrac{1}{\sqrt{3}}$

$t = \pm 27.36...°$

a) Show that Adam's solution is incorrect.

(1 mark)

b) Identify an error made by Adam.

....

....

(1 mark)

c) Bethan's teacher explains that one of her solutions is incorrect.
Identify and explain the error Bethan has made.

....

....

(2 marks)

7 The function $f(\theta) = \tan^2 \theta + \dfrac{\tan \theta}{\cos \theta}$ is defined for $0 \le \theta \le 2\pi$, $\theta \ne \dfrac{\pi}{2}, \dfrac{3\pi}{2}$.

a) Show that the equation $\tan^2 \theta + \dfrac{\tan \theta}{\cos \theta} = 1$ can be written in the form $2 \sin^2 \theta + \sin \theta - 1 = 0$.

(3 marks)

b) Hence find all solutions to the equation $f(\theta) = 1$ in the interval $0 \le \theta \le 2\pi$.

....

(4 marks)

Trigonometry

8   The diagram on the right shows the graph of $y = \arccos x$, where y is in radians. A and B are the end points of the graph.

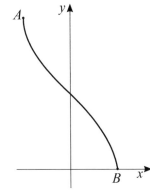

a)   Write down the coordinates of A and B.

$$A = \text{............................} \, , B = \text{............................}$$

(2 marks)

b)   Express x in terms of y.

....

(1 mark)

c)   Solve, to 3 significant figures, the equation $\arccos x = 2$ for the interval shown on the graph.

....

(2 marks)

9   In this question, give your answers to 3 significant figures where appropriate.

a)   Find the values of θ in the range $0 \le \theta \le 2\pi$ for which $\operatorname{cosec} \theta = \frac{5}{3}$.

....

(2 marks)

b)   **(i)**   Use an appropriate identity to show that $3 \operatorname{cosec} \theta = \cot^2 \theta - 17$ can be written as $18 + 3 \operatorname{cosec} \theta - \operatorname{cosec}^2 \theta = 0$.

(2 marks)

(ii)   Hence solve the equation $3 \operatorname{cosec} \theta = \cot^2 \theta - 17$ for $0 \le \theta \le 2\pi$.

....

(5 marks)

10   Given that θ is small and measured in radians, show that $4 \sin \theta \tan \theta + 2 \cos \theta$ can be approximated by the expression $p + q\theta^2$, where p and q are integers to be found.

(3 marks)

Trigonometry

11 Given that $\cos x = \dfrac{8}{9}$ for the acute angle x, find the exact values of:

a) $\sec x$

....

(1 mark)

b) $\operatorname{cosec} x$

....

(2 marks)

c) $\tan^2 x$

....

(2 marks)

d) $\cos 2x$

....

(3 marks)

12 The graph below shows the curve of $y = \dfrac{1 + \cos x}{2}$:

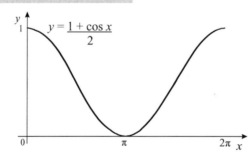

a) Show that $\dfrac{1 + \cos x}{2} = \cos^2 \dfrac{x}{2}$.

Use the double angle
formula for cos.

(2 marks)

b) Hence find the exact values of x for which $\cos^2 \dfrac{x}{2} = 0.75$ in the interval $0 \le x \le 2\pi$.

....

(2 marks)

Trigonometry

13 By writing $\sin 2\theta$ in terms of $\sin\theta$ and $\cos\theta$, solve the equation $3\sin 2\theta \tan\theta = 5$, for $0 \le \theta \le 2\pi$. Give your answers to 3 significant figures.

....

(6 marks)

14 Figure 1 shows an isosceles triangle ABC with $AB = AC = 2\sqrt{2}$ cm and $\angle BAC = 2\theta$. The midpoints of AB and AC are D and E respectively. A rectangle $DEFG$ is drawn inscribed in the triangle, with F and G on BC. The perimeter of rectangle $DEFG$ is P cm.

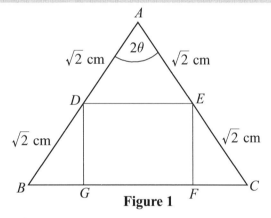

Figure 1

a) Using the cosine rule, $DE^2 = 4 - 4\cos 2\theta$.
Use a trigonometric identity to show that $DE = 2\sqrt{2}\,\sin\theta$.

(2 marks)

b) Show that $P = 4\sqrt{2}\,\sin\theta + 2\sqrt{2}\,\cos\theta$.

(2 marks)

c) Express P in the form $R\sin(\theta + \alpha)$ where $R > 0$ and $0 < \alpha < \dfrac{\pi}{2}$.

....

(3 marks)

Section One — Pure Maths

Trigonometry

15 A garden sprinkler system is set up to water a flower bed. The distance, d feet, the water sprays at time θ minutes can be modelled by the function $d = \sqrt{2}\cos\theta - 3\sin\theta$, where distance to the left of the sprinkler is modelled as negative and distance to the right of the sprinkler is modelled as positive.

a) Write $\sqrt{2}\cos\theta - 3\sin\theta$ in the form $R\cos(\theta + \alpha)$, where $R > 0$ and $0 \leq \alpha \leq \frac{\pi}{2}$.

....

(3 marks)

b) Find the times at which the water sprays 3 feet to the right of the sprinkler within the first 6 minutes after being switched on. Give your answers in minutes and seconds, to the nearest second.

....

(4 marks)

c) The distance (d) in feet that water is sprayed by an industrial sprinkler for a farmer's field is modelled by the function $d = (\sqrt{2}\cos\theta - 3\sin\theta)^4$. θ is the time in minutes. Find the maximum distances to the left and right of this sprinkler that the water reaches.

....

(2 marks)

d) Give one possible explanation for the minimum distance found in part c) in the context of this model.

....

....

(1 mark)

16 Show that $2\tan A \operatorname{cosec} 2A \equiv 1 + 1\tan^2 A$.

(3 marks)

Be very careful with degrees and radians. If the question gives the range in radians, you must give your answer in radians, otherwise you'll lose marks. Make sure your calculator is set to radians when appropriate too. There are some bits of trig that you have to use radians for — like the small angle approximations and arc length and sector area. Better get used to them.

Score

95

Exponentials and Logarithms

Exponentials and logs might seem a bit tricky at first, but once you get used to them and they get used to you, you'll wonder what you ever worried about. Plus, they're really handy for modelling real-life situations.

1   Given that $p > 0$, what is the value of $\log_p(p^4) + \log_p(\sqrt{p}) - \log_p\left(\dfrac{1}{\sqrt{p}}\right)$?

....

(3 marks)

2   Solve the equation $5^{(z^2 - 9)} = 2^{(z - 3)}$, giving your answers to 3 significant figures where appropriate.

....

(5 marks)

3   Solve the equation $3^{2x} - 9(3^x) + 14 = 0$, giving each solution to an appropriate degree of accuracy.

> This should remind you of a quadratic equation...

....

(5 marks)

4   The curve with equation $y = \ln(4x - 3)$ is shown on the graph to the right.

    a)   The point A with coordinates $(a, 1)$ lies on the curve.
             Find a to 2 decimal places.

$a = $ ....

(2 marks)

    b)   The curve only exists for $x > b$. State the value of b.

$b = $ ....

(2 marks)

Exponentials and Logarithms

5 The curve below has equation $y = Ae^{bx}$.

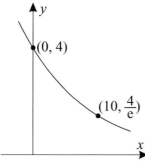

a) Find the values of A and b.

$A =$ .... $b =$ ....

(3 marks)

b) Find the exact coordinates of the point with gradient -1.

....

(5 marks)

6 The sketch on the right shows the function $y = e^{ax} + b$, where a and b are constants.

Find the values of a and b, and the equation of the asymptote shown on the sketch.

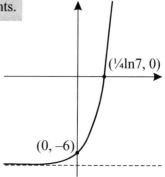

....

(4 marks)

7 Rudy buys a motorbike for £8000. The value of the motorbike, £m, after t months, is modelled by the function $m = m_0 e^{kt}$, where m_0 is the initial value of the motorbike.
k is given by the formula $k = \frac{1}{12} \ln\left(1 - \frac{r}{100}\right)$, where $r\%$ is the rate of depreciation per year.
For the first year, the motorbike depreciates by 8% per year, then the rate drops to 4% per year for each subsequent year. After how many months will the motorbike be worth less than half its original value?

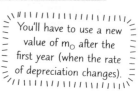

You'll have to use a new value of m_0 after the first year (when the rate of depreciation changes).

....

(5 marks)

Exponentials and Logarithms

8   The UK population, P, of an endangered species of bird is modelled over time, t years ($t \geq 0$), by the function: $P = 5700e^{-0.15t}$. The UK population, Q, of a bird of prey that hunts the endangered species, as well as other animals, is modelled by the function: $Q = 2100 - 1500e^{-0.15t}$. The time $t = 0$ represents the beginning of the year 2010, when the bird of prey was first introduced into the country.

a)   Find the year in which the population of the bird of prey is first predicted to exceed the population of the endangered species according to these models.

....

(4 marks)

b)   The graph showing the predicted population of the bird of prey is shown on the right. Add a curve to the graph to show the predicted UK population of the endangered species of bird over the same time period.

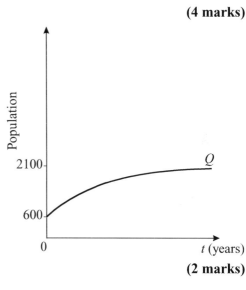

(2 marks)

c)   Comment on the validity of each population model.

....

....

....

....

(2 marks)

d)   Predict the year that the population of the endangered species will drop to below 1000.

....

(3 marks)

e)   When this population drops below 1000, conservationists start enacting a plan to save the species. Suggest one refinement that could be made to the model to take this into account.

....

....

(1 mark)

Exponentials and Logarithms

9   The number of supporters of a local football team has tended to increase in recent years. The attendance can be modelled by an equation of the form $y = ab^t$, where y is the average home game attendance in hundreds, t is the number of years after the 2010/11 season, and a and b are constants to be determined.

a)   Show that $y = ab^t$ can be written in the form $\log_{10}y = t\log_{10}b + \log_{10}a$.

(2 marks)

The graph of $\log_{10}y$ against t has been plotted below. A line of best fit has been drawn on the graph.

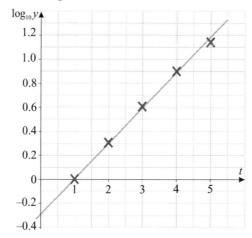

b)   A new stadium will be built for the team when the average home game attendance exceeds 5000. Use the information provided to predict the season in which the attendance will reach this value.

....

(5 marks)

c)   Interpret the value of a in the context of the model.

....

....

(1 mark)

d)   Can this model be used to accurately predict the number of fans in the 2024/25 season? Explain your answer.

....

....

(2 marks)

If you're asked to comment on the validity of models, or asked for limitations or refinements, think about anything that stops the model being realistic. For example, it might predict that a quantity will continue to rise until it becomes infinitely big, which might not be likely in a real-life situation — so you could refine the model by introducing an upper limit.

Score

56

Differentiation 1

Differentiation can tell you all sorts of useful things about the gradient of a curve. The following questions use simple functions in powers of x — you'll find trickier functions in 'Differentiation 2'.

1 The curve C is given by the equation $y = 2x^3 - 10x^2 - 4\sqrt{x} + 12$.

Find the gradient of the tangent to the curve at the point where $x = 4$.

....

(4 marks)

2 For $f(x) = x^3 - 7x^2 + 8x + 9$, sketch the graph of $y = f'(x)$, showing clearly the points of intersection with the axes.

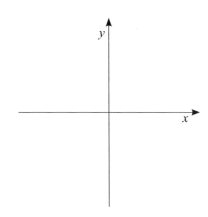

(4 marks)

3 A curve has equation $y = kx^2 - 8x - 5$, for a constant k. The point R lies on the curve and has an x-coordinate of 2. The normal to the curve at point R is parallel to the line with equation $4y + x = 24$.

a) Find the value of k.

$k = $ ....

(5 marks)

b) The tangent to the curve at R meets the curve $y = 4x - \dfrac{1}{x^3} - 9$ at the point S. Find the coordinates of S.

....

(5 marks)

Differentiation 1

4 For $f(x) = 8x^2 - 1$, prove from first principles that $f'(x) = 16x$.

(4 marks)

5 The function $f(x) = 2x^4 + 27x$ has one stationary point.

    a) Find the coordinates of the stationary point.

....

(4 marks)

    b) Find the range of values of x for which the function is increasing and the range of values of x for which it is decreasing.

Increasing for: ...., decreasing for: ....

(2 marks)

    c) Hence sketch the curve $y = f(x)$, showing where it crosses the axes and the position of its stationary point.

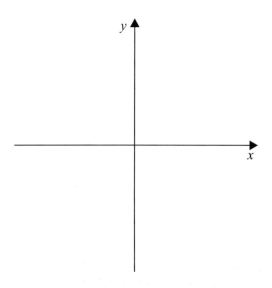

(3 marks)

Differentiation 1

6 The diagram shows part of the graph of $y = x^4 + 3x^3 - 6x^2$.
Find the range of values of x for which the graph is concave.

Start by differentiating twice to find $\frac{d^2y}{dx^2}$.

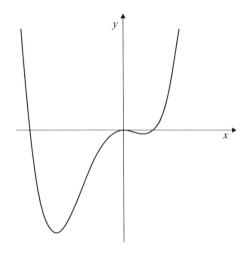

....

(4 marks)

7 The function $f(x) = 3x^3 + 9x^2 + 25x$ has one point of inflection.

a) Show that the point of inflection is at $x = -1$.

(5 marks)

b) Explain whether or not this point of inflection is a stationary point.

....

....

(2 marks)

c) Joe claims that the function $f(x) = 3x^3 + 9x^2 + 25x$ is an increasing function for all values of x.
Show that Joe is correct.

(2 marks)

Differentiation 1

8 An ice cream parlour needs an open-top stainless steel container with a capacity of 40 litres, modelled as a cuboid with sides of length x cm, x cm and y cm, as shown in Figure 1.

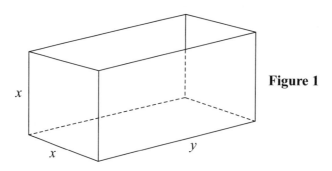

Figure 1

a) Show that the external surface area, A cm^2, of the container is given by $A = 2x^2 + \dfrac{120\,000}{x}$.

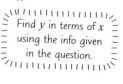

Find y in terms of x using the info given in the question.

(4 marks)

b) Find the value of x to 3 s.f. at which A is stationary, and show that this is a minimum value of A.

(6 marks)

c) Calculate the minimum area of stainless steel needed to make the container. Give your answer to 3 s.f.

.... cm^2

(2 marks)

d) Comment on the validity of this model.

....

....

(1 mark)

EXAM TIP Remember, tangents have the same gradient as the curve and normals are perpendicular to the curve. To find stationary points, differentiate the expression and set equal to zero. And one more thing... f''(x) > 0 means the gradient is increasing (the curve is convex), f''(x) < 0 means the gradient is decreasing (the curve is concave) and f''(x) = 0 at a point of inflection. Phew.

Score

57

Differentiation 2

As promised, more differentiation. Make sure you use the correct rule for each type of function and you'll be fine.

1   Find $\dfrac{dy}{dx}$ at the given point for each of the following:

    a) $y = \dfrac{1}{\sqrt{2x - x^2}}$, $(1, 1)$

....
(4 marks)

    b) $x = (4y + 10)^3$, $(8, -2)$

....
(4 marks)

2   The diagram shows a container in the shape of a regular tetrahedron, which is being filled with water. After t minutes, the water in the container can be modelled as a regular tetrahedron with edge length a cm and vertical height x cm, and the volume of water in the container is V cm^3.

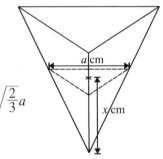

    a)   Given that a regular tetrahedron with edge length a has vertical height $h = \sqrt{\dfrac{2}{3}}\,a$ and volume $\dfrac{\sqrt{2}}{12}a^3$, show that the volume of water in the container after t minutes is given by $V = \dfrac{\sqrt{3}}{8}x^3$.

(2 marks)

Water is being poured into the container at a constant rate of 240 cm^3 min^{-1}.

    b)   Find the exact value of $\dfrac{dx}{dt}$ when the vertical height of the water in the container is 8 cm.

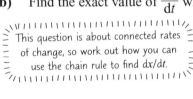
This question is about connected rates of change, so work out how you can use the chain rule to find dx/dt.

....
(5 marks)

The value of $\dfrac{dx}{dt}$ is measured when $x = 12$ and found to be $\dfrac{32}{9\sqrt{3}}$. This is less than the value expected if $\dfrac{dV}{dt} = 240$ cm^3 min^{-1}. It is discovered that water has been leaking out of the container at a constant rate.

    c)   Find $\dfrac{dV}{dt}$ if $\dfrac{dx}{dt} = \dfrac{32}{9\sqrt{3}}$ when $x = 12$.

....
(3 marks)

Differentiation 2

3 A curve has the equation $y = e^{2x} - 5e^x + 3x$.

    a) Find $\dfrac{dy}{dx}$.

....
(2 marks)

    b) Show that the stationary points on the curve occur when $x = 0$ and $x = \ln \dfrac{3}{2}$.

(4 marks)

    c) Determine the nature of each of the stationary points.

(3 marks)

4 The curve C is given by the equation $y = 2^{3x}$.

    a) Find the exact value of $\dfrac{dy}{dx}$ at the point on the curve where $x = 1$.

....
(3 marks)

    b) Explain why $y = 2^{3x}$ is an increasing function for all values of x.

....

....

....
(2 marks)

Differentiation 2

5 Using differentiation from first principles, find the exact value of the derivative of $\cos x$ at the point $x = \frac{\pi}{3}$.

You'll need to use the addition formulas and small angle approximations here.

....

(6 marks)

6 For $y = \ln x\,(5x-2)^3$, show that $\dfrac{dy}{dx} = (5x-2)^2\left(a\ln x + b + \dfrac{c}{x}\right)$, where a, b and c are constants to be found.

(4 marks)

7 A curve has the equation $y = \dfrac{5 - 2x}{3x^2 + 3x}$.

Show that, at the stationary points on the curve, $2x^2 - 10x - 5 = 0$.

(6 marks)

8 A curve has the equation $y = 4x^2 \ln x, \quad x > 0$.

Find the ranges of values of x for which the curve is concave and convex.

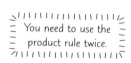

You need to use the product rule twice.

Concave for: ...., Convex for: ....

(8 marks)

Section One — Pure Maths

Differentiation 2

9 The diagram shows part of the curve $y = \dfrac{4x - 1}{\tan x}$, $0 < x < \pi$.

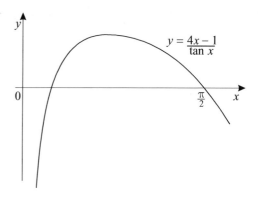

a) Show that an expression for $\dfrac{dy}{dx}$ is:

$$\frac{dy}{dx} = 4 \cot x - (4x - 1) \operatorname{cosec}^2 x$$

(3 marks)

b) The curve has a maximum in the range $0 < x < \dfrac{\pi}{2}$.

Show that at the maximum point, $2 \sin 2x - 4x + 1 = 0$.

(4 marks)

10 A curve is defined by the parametric equations $x = t^2 + 1$, $y = t^3 + 2t$.

Find the equation of the tangent to the curve at the point where $t = 1$.
Give your answer in the form $y = mx + c$.

....

(5 marks)

Differentiation 2

11 For $y = \cos^{-1} x$, show that $\dfrac{dy}{dx} = -\dfrac{1}{\sqrt{1-x^2}}$.

(4 marks)

12 The curve C is defined by the parametric equations $x = \dfrac{\sin\theta}{2} - 3$, $y = 5 - \cos 2\theta$.

a) Find an expression for $\dfrac{dy}{dx}$ in terms of θ.

....

(2 marks)

b) Hence find the equation of the tangent to C at the point where $\theta = \dfrac{\pi}{6}$.
Give your answer in the form $ax + by + c = 0$.

....

(3 marks)

c) The line $y = -8x - 20$ crosses C at point P. Find the coordinates of P, showing your working clearly.

....

(6 marks)

d) Find a Cartesian equation for C in the form $y = f(x)$.

....

(3 marks)

Section One — Pure Maths

Differentiation 2

13 A curve has the equation $x^3 + x^2y = y^2 - 1$.

    a) Find an expression for $\dfrac{dy}{dx}$.

....
(4 marks)

The points P and Q lie on the curve. P has coordinates $(1, a)$ and Q has coordinates $(1, b)$.

    b) Find the values of a and b, given that $a > b$.

....
(2 marks)

    c) Find the equation of the normal to the curve at Q.
        Give your answer in the form $y = mx + c$.

....
(3 marks)

14 A curve has the equation $\sin \pi x - \cos\left(\dfrac{\pi y}{2}\right) = 0.5$, for $0 \le x \le 2$, $0 \le y \le 2$.

The curve has one stationary point. Find the coordinates of the stationary point.

....
(7 marks)

Score

102

Integration 1

Integration starts off pretty easy, but then gets really hard really quickly. Just keep your wits about you, and watch out for places where you can use clever tricks (like when the numerator is the derivative of the denominator).

1 Find $\int \left(\dfrac{x^2+3}{\sqrt{x}} \right) dx$.

....

(3 marks)

2 The curve C has the equation $y = f(x)$, $x > 0$. $f'(x)$ is given as $2x + 5\sqrt{x} + \dfrac{6}{x^2}$.

A point P on curve C has the coordinates $(3, 7)$. Find $f(x)$, giving your answer in its simplest form.

....

(6 marks)

3 Region A is bounded by the curve $y = \dfrac{2}{\sqrt{x^3}}$ ($x > 0$), the x-axis and the lines $x = 2$ and $x = 4$.

Show that the area of A is $2\sqrt{2} - 2$.

(5 marks)

Integration 1

4   Evaluate $\int_{p}^{4p} \left(\dfrac{1}{\sqrt{x}} - 4x^3 \right) dx$, where $p > 0$, leaving your answer in terms of p.

(4 marks)

5   The curve $y = 2x^3 - 3x^2 - 11x + 6$ is shown below. It crosses the x-axis at $(-2, 0)$, $(0.5, 0)$ and $(3, 0)$.

Find the area of the shaded region bounded by the curve, the x-axis and the lines $x = -1$ and $x = 2$.

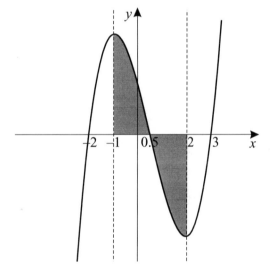

(6 marks)

6   Find the possible values of k that satisfy $\int_{\sqrt{2}}^{2} (8x^3 - 2kx) \, dx = 2k^2$, where k is a constant.

(5 marks)

Integration 1

7   The graph on the right shows the curves $y = \dfrac{8}{x^2}$ and $y = 9 - x^2$ for $x, y \geq 0$.

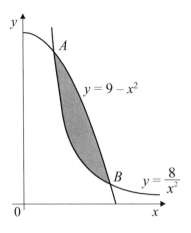

    a)   Find the coordinates of points A and B.

....

(5 marks)

    b)   Find the shaded area. Give your answer in the form $a + b\sqrt{2}$, where a and b are fractions.

....

(4 marks)

8   The graph below shows the curve C, which has equation $y = \sqrt{x} - \dfrac{1}{2}x^2 + 1 \ (x \geq 0)$.

Point M lies on the curve and has coordinates $\left(1, \dfrac{3}{2}\right)$, and line N is the normal to the curve at point M.
The shaded region A is bounded by the y-axis, the curve and the normal to the curve at M.

Find the area of A.

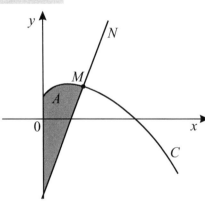

> Combine the equations of the curve and the line to find the function to integrate.

....

(9 marks)

Integration 1

9 The graph below shows the curve C, which has parametric equations $x = 8t^3$ and $y = 2t - 16t^4$ (for $t \geq 0$).
The curve intersects the x-axis at the point $(1, 0)$.
The shaded region A is bounded by the curve C and the x-axis.

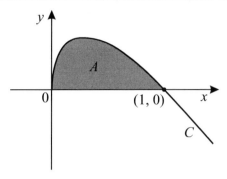

a) Find t when $x = 0$ and $x = 1$.

....
(1 mark)

b) Hence, or otherwise, find the area of A.
Give your answer as a fraction in its simplest form.

Use the chain rule to integrate parametric equations — and don't forget to change the limits.

....
(4 marks)

10 Find the exact value of p given that $\displaystyle\int_{2p}^{6p} \frac{x^3 + 4x^2}{x^3}\, dx = 4 \ln 12$.

$p = $ ....
(4 marks)

Integration 1

11 Find $\int \dfrac{1}{x(3x-2)}\,dx$.

....
(5 marks)

12 $f(x) = \dfrac{3x+5}{(3x+1)(1-x)}$

    a) Write $f(x)$ in partial fractions.

....
(3 marks)

    b) Find the exact value of $\int_{-2}^{4} f(x)\,dx$ in the form $\ln \dfrac{p}{q}$, where p and q are integers.

....
(3 marks)

13 Find the exact value of $\int_{-1}^{1} \dfrac{4x-10}{4x^2+4x-3}\,dx$, giving your answer in the form $\ln k$, where k is an integer.

> Start by factorising
> the denominator here.

....
(7 marks)

EXAM TIP Make sure you're completely happy with integrating powers — including the ones that give you a natural log (ln) when you integrate. It's usually best to write roots and fractions as powers of x — it makes it much easier to integrate (and means you're less likely to make a mistake). To find tricky areas, try splitting them into different bits and finding each bit separately.

Score

74

Integration 2

As is typical with Hollywood blockbusters, after the box office success of Integration 1, here comes the sequel — Integration 2: Integration With A Vengeance. Just remember, in calculus, no one can hear you scream...

1   Find the exact value of $\int_{\frac{\pi}{12}}^{\frac{\pi}{8}} \sin 2x \, dx$.

(3 marks)

2   Find $\int \left(3 \tan^2 \left(\frac{x}{2} \right) + 3 \right) dx$.

(3 marks)

3   Find $\int \sec^2 x \, e^{\tan x} \, dx$.

(2 marks)

4   Find the exact value of $\int_0^{\frac{1}{2}} \frac{x}{1-x^2} \, dx$ using the substitution $x = \sin \theta$.

Don't forget to change the limits of integration when you make the substitution.

(7 marks)

Integration 2

5   Find $\int_{1}^{2}\left(\dfrac{\ln x}{\sqrt{x}}\right)^{2}dx$, using the substitution $u = \ln x$. Give your answer to 3 significant figures.

....
(5 marks)

6   Find $\int 4x e^{-2x}\,dx$.

> Use integration by
> parts for questions 6-8.

....
(4 marks)

7   Find:

a) $\int x \sin 4x\,dx$

....
(4 marks)

b) $\int x^{2}\cos 4x\,dx$

....
(4 marks)

8   Calculate the exact value of $\int_{1}^{4}\dfrac{\ln x}{2x^{2}}\,dx$.

....
(6 marks)

Integration 2

9 An ecologist is monitoring the population of newts in a colony. The rate of increase of the population is directly proportional to the square root of the current number of newts in the colony. When there were 36 newts in the colony, the rate of change was calculated to be 0.36.

a) Formulate a differential equation to model the rate of change, in terms of the variables N (number of newts), t (time in weeks).

....

(4 marks)

b) After more research, the ecologist decides that the differential equation

$\frac{\mathrm{d}N}{\mathrm{d}t} = \frac{kN}{\sqrt{t}}$, for a positive constant k, is a better model for the population.

When the ecologist began the survey, the initial population of newts in the colony was 25.

(i) Solve the differential equation, leaving your answer in terms of k and t.

....

(3 marks)

(ii) Given that the value of k is 0.05, calculate how long (to the nearest week) it will take for the population to double.

....

(3 marks)

10 A supermarket sets up an advertising campaign to increase sales on the cheese counter. After the start of the campaign, the number of kilograms of cheese sold each day, S, increases over time, t days. The increase in sales is modelled by the differential equation $\frac{\mathrm{d}S}{\mathrm{d}t} = k\sqrt{S}$ $(k > 0)$.

a) At the start of the campaign, the supermarket was selling 81 kg of cheese a day. Use this information to solve the differential equation, giving S in terms of k and t.

....

(3 marks)

b) Given that $\frac{\mathrm{d}S}{\mathrm{d}t} = 18$ at the start of the campaign, calculate the number of kg sold on the fifth day after the start of the campaign ($t = 5$).

....

(3 marks)

c) How many days will it take before the sales reach 225 kg a day?

....

(2 marks)

Integration 2

11 A solid hemisphere of radius r cm and surface area S cm^2 is decreasing in size.

a) The rate of decrease of r, over time t minutes, is directly proportional to rt.
   (i) Formulate a differential equation in terms of r, t and a positive constant k.

....

(2 marks)

   (ii) Show that $\dfrac{\mathrm{d}S}{\mathrm{d}t} = -2ktS$.

(4 marks)

b) The total surface area of the hemisphere is 200 cm^2 at time $t = 10$ minutes
   and 50 cm^2 at time $t = 30$ mins.
   (i) Find the particular solution of the equation $\dfrac{\mathrm{d}S}{\mathrm{d}t} = -2ktS$.
   Give the values of any constants to 3 significant figures.

....

(5 marks)

   (ii) Hence find the initial surface area of the hemisphere to 3 significant figures.

....

(1 mark)

c) A solid sphere with the same initial radius as the hemisphere is also decreasing in size.
   Maddy decides to use the differential equation given in part a) (ii) to model
   the surface area of the sphere. Explain why this will not be appropriate.

....

....

(1 mark)

Numerical Methods

There are two main parts to numerical methods — finding roots using iteration (including my personal favourite, the Newton-Raphson method) and numerical integration (which is also pretty darn exciting if you ask me).

1 The graph below shows the function $f(x) = 4(x^2 - 1)$, $x \geq 0$, and its inverse function $f^{-1}(x)$.

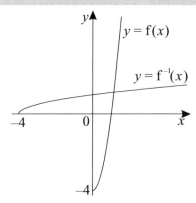

a) By finding an expression for $f^{-1}(x)$ and considering how the graphs are related, show that $\sqrt{\frac{x}{4} + 1} - x = 0$ at the point where the graphs meet.

(4 marks)

b) Show that the equation $\sqrt{\frac{x}{4} + 1} - x = 0$ has a root in the interval $1 < x < 2$.

(2 marks)

c) Starting with $x_0 = 1$, use the iteration formula:
$$x_{n+1} = \sqrt{\frac{x_n}{4} + 1}$$
to find the x-coordinate of the point of intersection, correct to 3 significant figures.

....

(3 marks)

d) Does the equation from part a) have any other roots? Explain your answer.

$\rightsquigarrow$ Look at the graph above...

....

....

(1 mark)

Numerical Methods

2 The sketch below shows the intersection of the curve $y = 6^x$ with the line $y = x + 2$ at the point P.

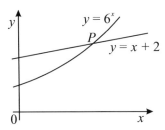

a) Show that a root of the equation $6^x - x - 2 = 0$ lies in the interval $[0.5, 1]$.

(2 marks)

b) Show that the Newton-Raphson iteration formula for finding the x-coordinate of P can be written as:

$$x_{n+1} = \frac{6^{x_n}(x_n \ln 6 - 1) + 2}{6^{x_n} \ln 6 - 1}$$

Remember — the derivative of a^x is $a^x \ln a$.

(4 marks)

c) Using the Newton-Raphson formula with a starting value of $x_0 = 0.5$, find the x-coordinate of P correct to 4 significant figures.

....
(3 marks)

d) By considering an appropriate interval, verify that this value is accurate to 4 significant figures.

(3 marks)

e) k is chosen as a different starting value. The gradient of the tangent at $x = k$ is 0.
Explain why the Newton-Raphson method will fail for this starting value.

....

....
(1 mark)

Numerical Methods

3 $f(x) = x^3 - x^2 + 4$

a) The equation $f(x) = 0$ has a root in the interval $(-2, -1)$.
For the starting value $x_0 = -1.5$, $f(x_0) = -1.625$ and $f'(x_0) = 9.75$.
Use the Newton-Raphson method to obtain a value for x_1, the second approximation for the root.
Give your answer to 4 significant figures.

....

(2 marks)

b) Show that the root, $b = -1.315$, is correct to 4 significant figures.

(3 marks)

c) Explain why the Newton-Raphson method fails when $x_0 = \frac{2}{3}$.

....

....

(2 marks)

4 The diagram on the right shows the graph of $y = 2^{x^2}$.

a) Use the trapezium rule with 4 intervals to find an estimate for the area of the region bounded by the axes, the curve and the line $x = 2$.
Give your answer to 3 significant figures.

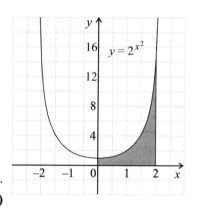

....

(4 marks)

b) Given that the curve is convex for all values of x, explain whether the estimate in a) is an overestimate or an underestimate.

....

....

(1 mark)

c) Suggest one way to find a more accurate estimate for this area using the trapezium rule.

....

....

(1 mark)

Numerical Methods

5   Figure 1 shows the graph of $y = x \sin x$. The region R is bounded by the curve and the x-axis ($0 \leq x \leq \pi$).

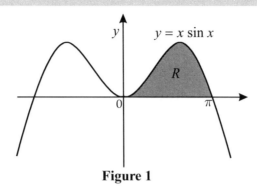

Figure 1

a)   Fill in the missing values of y in the table below.
Give your answers to 5 significant figures.

x	0	$\dfrac{\pi}{4}$	$\dfrac{\pi}{2}$	$\dfrac{3\pi}{4}$	π
y	0	0.55536			0

(1 mark)

b)   Hence find an approximation for the area of R, using the trapezium rule and all the values in the table.
Give your answer to 4 significant figures.

....
(3 marks)

c)   Find the exact area of R using integration by parts.

....
(6 marks)

d)   Hence find the percentage error of the approximation found in part b).
Give your answer to 2 significant figures.

Leave the answer from part b) in your calculator to get a more accurate value.

....
(2 marks)

Numerical Methods

6 Use the trapezium rule with 6 ordinates to estimate $\int_{1.5}^{4}(3x-\sqrt{2^x})\,dx$.
Give your answer to 4 significant figures.

....

(4 marks)

7 The graph below shows the curve $y=\dfrac{3\ln x}{x^2}$, $x>0$.
The shaded region R is bounded by the curve, the line $x=2.5$ and the line $x=3$.

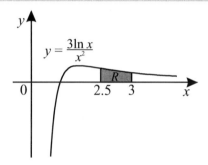

a) Find an approximation for the area of R, using the trapezium rule with 5 intervals.
Give your answer to 4 significant figures.

....

(4 marks)

b) By considering the areas of appropriate rectangles, show that the area of R is 0.2,
correct to 1 decimal place.

You need one rectangle that sits above the curve and one that sits below it.

(3 marks)

Score

59

Vectors

Aaah, good old dependable vectors. They've got a certain magnitude about them, and they always have a clear direction. Much like an inspirational leader. It's kind of how I see myself after I've brought about the revolution.

1 A vector of magnitude 7 acts vertically upwards. Another vector, of magnitude $4\sqrt{2}$, acts at an angle of 45° below the positive horizontal direction.
The resultant of the two vectors is **r**.

    a) Find **r** in terms of **i** and **j**.

$$\mathbf{r} = \text{..}$$
(2 marks)

    b) The vector **s** acts parallel to **r**, and has magnitude 35.
        Find **s** in terms of **i** and **j**.

$$\mathbf{s} = \text{..}$$
(3 marks)

2 Points A, B and C have position vectors $-\mathbf{i} + 7\mathbf{j} - 2\mathbf{k}$, $5\mathbf{i} - 3\mathbf{j} + 6\mathbf{k}$ and $5\mathbf{i} + 4\mathbf{j} + 3\mathbf{k}$ respectively.
M is the midpoint of AB.

Find the exact value of k such that $|\overrightarrow{CM}| = k\,|\overrightarrow{AB}|$.

$$k = \text{................................}$$
(5 marks)

3 Given that $\overrightarrow{OA} = -2\mathbf{i} + 4\mathbf{j} - 5\mathbf{k}$, $\overrightarrow{OB} = 14\mathbf{i} + 12\mathbf{j} - 9\mathbf{k}$ and $\overrightarrow{OC} = 2\mathbf{i} + \mu\mathbf{j} + \lambda\mathbf{k}$, find the values of μ and λ such that points A, B and C are collinear.

If two vectors are parallel and share a point, they lie on the same straight line.

$$\mu = \text{...............} \qquad \lambda = \text{...............}$$
(5 marks)

Section One — Pure Maths

Vectors

4   Figure 1 shows a sketch of a triangle, PQR. Given that $\overrightarrow{PQ} = \begin{pmatrix} 2 \\ -9 \end{pmatrix}$ and $\overrightarrow{QR} = \begin{pmatrix} 14 \\ 6 \end{pmatrix}$, find the angle at P.

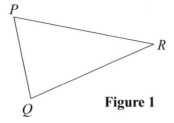

Figure 1

.... °

(5 marks)

5   Points A, B and C have position vectors $\begin{pmatrix} 1 \\ -3 \\ 2 \end{pmatrix}$, $\begin{pmatrix} 4 \\ -12 \\ 8 \end{pmatrix}$ and $\begin{pmatrix} -3 \\ 9 \\ -6 \end{pmatrix}$ respectively.

Point D lies on AB such that $AD:DB = 2:1$.

Point E is positioned such that $\overrightarrow{OD} = -\frac{1}{2}\overrightarrow{CE}$.

Find the position vector of point E.

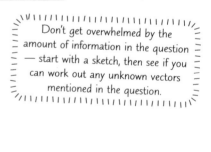

Don't get overwhelmed by the amount of information in the question — start with a sketch, then see if you can work out any unknown vectors mentioned in the question.

....

(5 marks)

Vectors

6 Two drones, A and B, take off from a launch pad at origin O and follow a series of movements based on the vectors $\mathbf{a} = (4\mathbf{i} + 6\mathbf{j} + 5\mathbf{k})$ metres, $\mathbf{b} = (-\mathbf{i} - 2\mathbf{j} - 2\mathbf{k})$ metres and $\mathbf{c} = (-3\mathbf{j} + \mathbf{k})$ metres, where $\mathbf{i}$ and $\mathbf{j}$ are horizontal unit vectors and $\mathbf{k}$ is a vertical unit vector with the upwards direction being positive.

Drone A follows the vectors $\mathbf{a}$, then $\mathbf{b}$, then $\mathbf{c}$, and then maintains its final position.
Drone B follows the vectors $2\mathbf{a}$, then $\mathbf{b}$, then $-3\mathbf{c}$, and then maintains its final position.

a) Calculate to 3 s.f. the distance between the final positions of drones A and B.

.... m

(4 marks)

b) Drone A now moves in a straight line to a position exactly 2 m directly below drone B.
Find the vector that describes this movement of drone A.

....

(2 marks)

c) To maintain contact with the controller on the ground, drone B must always stay within 50 m of the origin O. How far can drone B move in the positive $\mathbf{j}$ direction from its current position before it passes out of range? Give your answer to 3 s.f.

.... m

(5 marks)

EXAM TIP Fun, fun, fun, vectors are just pure fun, don't you think... ... ... Well anyway, make sure you're comfortable finding their magnitudes and using trig to find angles. Don't be put off by 3D ones — they're generally pretty much the same as their 2D counterparts. Some questions can be quite involved, but always start with a sketch, take it step by step and you'll be OK.

Score

36

Data Presentation and Interpretation

Questions on presenting and interpreting data are the perfect chance for the examiners to test you on the large data set — so now would be a good time to have another look at those massive spreadsheets...

1   A group of 10 friends play a round of mini golf and record their scores, x.
It is given that $\sum x = 500$ and $\sum x^2 = 25\,622$.

   a)   Find the mean and the standard deviation for the data.

mean = ...,   standard deviation = ...

(3 marks)

   b)   Another friend wants to incorporate his score of 50. Giving reasons, but without further calculation, explain the effect of adding this score on the mean.

....

....

(2 marks)

   c)   A second group of friends record their mini golf scores on the same course. The mean of these scores is 58 and the standard deviation is 4.13. Compare the sets of scores for the two groups.

....

....

(2 marks)

2   Norman is investigating the methods of transport used by the people in one Local Authority to get to their place of work. He summarises data from the 2001 and 2011 UK censuses to create the table below.
He gives each method of transport his own rating for 'exercise done by the user', from 1 (low) to 4 (high).

	Method of transport used to get to work	Private motorised	Public motorised	On foot	Bicycle
	Exercise rating	1	2	3	4
2001 census	Number of people	55 170	25 697	11 235	1781
	Percentage	58.8	27.4	12.0	1.9
2011 census	Number of people	62 253	29 413	15 302	3223
	Percentage	56.5	26.7	13.9	2.9

   a)   Norman claims that the exercise ratings were higher overall in 2011 than in 2001.
Without doing any calculations, use data from the table to support Norman's claim.

....

(1 mark)

   b)   Norman has given a higher exercise rating to 'On foot' than 'Public motorised'.
Comment on whether this is appropriate for this data set.

....

....

(1 mark)

Data Presentation and Interpretation

3 All the Year 12 students in a school were asked how much their monthly phone bill is (in £). The incomplete table and histogram below show the results.

a) Complete the table and the histogram.

Monthly phone bill, £b	Frequency
$0 \leq b < 10$	12
$10 \leq b < 15$	23
$15 \leq b < 18$	
$18 \leq b < 20$	
$20 \leq b < 25$	18
$25 \leq b < 35$	6

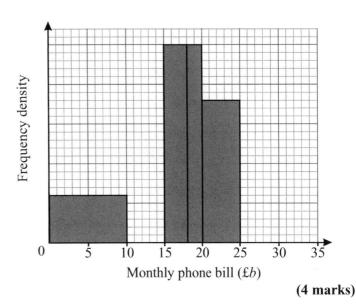

(4 marks)

b) Estimate the number of students that have a monthly phone bill of between £12.50 and £17.50.

....

(2 marks)

c) Estimate the mean monthly phone bill for the Year 12 students.

£ ....

(2 marks)

d) Estimate the standard deviation for the data.

£ ....

(2 marks)

e) Estimate the 20th percentile of the data.

£ ....

(3 marks)

Data Presentation and Interpretation

4   The percentages of people in different age groups for Eastbourne and Camden in 2011 are shown on this chart.

- In 2011, the mean age for Eastbourne was 42.8 years and the standard deviation was estimated to be 23.7 years (3 s.f.).

- In 2011, the mean age for Camden was 35.7 years and the standard deviation was estimated to be 19.1 years (3 s.f.).

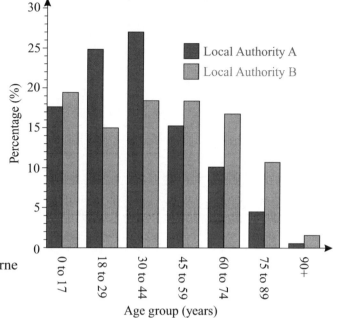

a)   Match each of the Local Authorities (Eastbourne and Camden) to the corresponding bars on the graph. You must explain your reasoning.

Local Authority A = ...

Local Authority B = ...

Reasoning: ....

....

....

(2 marks)

b)   Using an appropriate calculation to support your answer, explain why the Eastbourne data for 2011 is likely to have outliers at the upper end of the data.

....

....

(2 marks)

c)   In 2011, there were 220 338 people living in Camden. Some data values are classed as outliers. Suggest one reason why the mean might still be an appropriate measure of location for the 2011 Camden data.

....

....

(1 mark)

Data Presentation and Interpretation

5   The heights of giraffes living in a zoo were measured.
     The results are shown on the cumulative frequency diagram below.

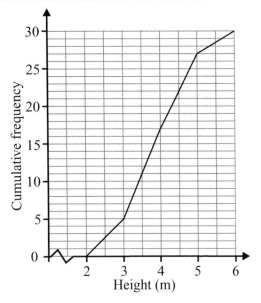

a)   Use the cumulative frequency diagram to estimate the median, lower quartile and upper quartile
      of the data. Then calculate an estimate for the interquartile range.

median = ... m,   lower quartile = ... m,

upper quartile = ... m,   interquartile range = ... m

(4 marks)

The tallest giraffe in the zoo measures 5.56 m and the shortest giraffe measures 2.7 m.

The heights of giraffes living in a game reserve were also measured.
This data is summarised in the box plot below.

Height (m)

b)   Compare the heights of the two groups of giraffes.

....

....

....

(3 marks)

Data Presentation and Interpretation

6 The sales figures for a gift shop over a 12-week period are shown on the stem and leaf diagram below.

```
3 | 8
4 | 1 2 6 9
5 | 5 8 9
6 | 0 2 4
7 |
8 |
9 | 1
```

Key
3 | 8 means £3800

a) Explain why the stem and leaf diagram is a suitable way to display this data.

....

....

(1 mark)

b) Find the median, lower quartile, upper quartile and interquartile range of the sales data.

median = £ ....,   lower quartile = £ ....,

upper quartile = £ ....,   interquartile range = £ ....

(4 marks)

The shop's manager is considering excluding any outliers from his analysis of the data,
to get a more realistic idea of how the shop is performing.

c) Explain, using a calculation, why the data value £9100 could be considered an outlier.

....

....

(1 mark)

d) Do you think the manager should include this outlier in his analysis? Explain your answer.

....

....

(1 mark)

e) The mean and standard deviation for this sales data are calculated to be: mean = £5540,
standard deviation = £1370 (both to 3 s.f.). Do you think these two measures, or the median and
interquartile range, are more useful measures of location and spread for this data? Explain your answer.

....

....

....

(2 marks)

Data Presentation and Interpretation

7 The graph below shows some data about the times taken for runners to finish the Alverston marathon. The graph shows the mean time taken, in minutes, by different age groups of runners who ran the marathon in 2000, 2005, 2010 and 2015.

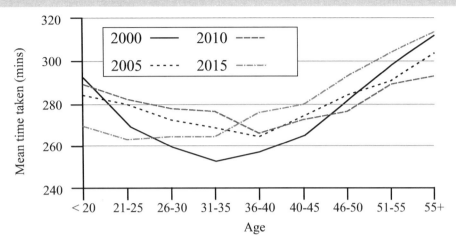

a) State the mean time taken for people aged 40-45 years to finish the marathon in 2015.

.... minutes

(1 mark)

b) A researcher studies a randomly-chosen sample of runners who were in the 21-25 age group in 2000. Does the graph suggest these runners' times improved in 2010? Justify your answer.

....

....

....

(2 marks)

c) Write down one assumption you have made about the runners in your answer to part b).

....

....

(1 mark)

Serj wants to find the overall mean time for the marathon runners in 2015. To calculate this, he intends to add up the mean times for the 9 individual age categories, then divide by 9.

d) Explain why Serj's calculation is unlikely to be accurate, and suggest how he could improve his method.

....

....

....

(2 marks)

As well as being able to calculate measures like the mean and the standard deviation, it's important to know what they tell you about a data set. Once you understand what they are and why they're useful, questions asking you to interpret or discuss data become a breeze — which means picking up those extra marks in the exam without needing to do any tricky maths.

Score

49

Probability

Venn diagrams. My question is, Vhat about the Vhere, the Vhich and the Vhy diagrams? I'm sorry, that isn't even remotely funny. In fact, I often find that me making jokes and people laughing are mutually exclusive events...

1   The events A and B are mutually exclusive. P(A) = 0.1 and P(B) = 0.4.
    Event C has probability P(C) = 0.3.
    Events B and C are independent, and the probability
    of both events A and C occurring is 0.06.

   a)   Draw a Venn diagram showing the probabilities of events A, B and C.

                                                                              (5 marks)

   b)   Are events A and C independent? Explain your answer.

        ....

        ....
                                                                              (2 marks)

   c)   Find P(B ∪ C)

                                                                    ....
                                                                              (1 mark)

   d)   Find P(A' ∩ B')

                                                                    ....
                                                                              (1 mark)

   e)   Find P(B'|A'), to 2 decimal places.

                                                                    ....
                                                                              (2 marks)

Probability

2 This incomplete Venn diagram shows the probabilities of two independent events L and M. Calculate P(L' ∩ M').

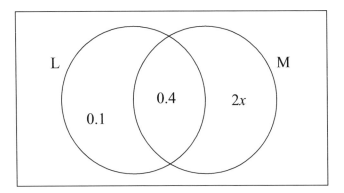

....

(4 marks)

3 Of 30 drivers interviewed, 9 have been involved in a car crash in the last year. Of those who have been involved in a crash, 5 wear glasses. The probability of wearing glasses, given that the driver has not been involved in a car crash is $\frac{1}{3}$.

Find the probability that a glasses-wearer interviewed has been involved in a car crash.

....

(4 marks)

Probability

4 Alan is studying the age of people attending Ulverston Dynamo's home matches. He surveyed every fan attending the match against Cark Sportif, and recorded their ages and gender in the table shown below.

Age Group	Frequency	
	Male	Female
Under 18	305	207
18-30	431	333
31-45	378	531
46-60	472	236
Over 60	359	115

a) Find the probability that a fan selected at random from the crowd is under 18 given that they are male.

....

(2 marks)

b) Alan uses his data to model probabilities for the crowd at Ulverston Dynamo's next home match, which is against AC Flookburgh. He says the probability that a fan selected at random from the crowd is in the 18-30 age group is $\frac{764}{3367}$. Describe one assumption Alan has made in his model, and explain why this assumption may not be valid.

....

....

(2 marks)

5 Jessica has two packs of cards. One pack has some cards missing. The probability of selecting a heart from this pack, P(S) is 0.3. Jessica selects one card from this pack and one card from a complete pack of 52 cards.

a) Find the probability that at least one of the cards is a heart.

....

(2 marks)

Jessica then selects three cards from another complete pack without replacement.

b) Find the probability that she selects at least two hearts, given that the first card she selected was a heart. Give your answer to 3 significant figures.

Without replacement means there'll be a different number of cards each time she selects one — and don't forget about the conditional part too.

....

(3 marks)

c) Give one assumption you made in answering parts a) and b).

....

....

(1 mark)

Probability

6 P(A | B) = 0.31, P(B | A) = 0.25 and P(B) = 0.4. Find P(A').

....
(3 marks)

7 At a school cafeteria, every child takes at least one type of fruit with their lunch.
The cafeteria has bananas, apples and oranges.

65 children use the cafeteria.
25 take a banana, 28 take an orange, 32 take an apple, 5 take a banana and an orange,
6 take an orange and an apple, and 4 take an orange, an apple and a banana.

a) Find the number of children who take a banana and an apple.

....
(4 marks)

Two children are picked at random, one after the other, without replacement.

b) Given that at least one child takes an apple, find the exact probability that the first child took an apple.

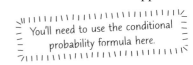
You'll need to use the conditional
probability formula here.

....
(4 marks)

EXAM TIP

Even if the question doesn't ask you to draw a Venn or tree diagram, it's often helpful to draw one. They make your life so much easier. Also, make sure you understand all the different bits of set notation — it'd be silly to get a question wrong because you mixed up the union (∪) and intersection (∩) symbols. Just remember, 'u for **u**nion and n for i**n**tersection'.

Score

40

Statistical Distributions

Another treat for probability fans (like me), and your first glimpse of the binomial and normal distributions.
If you don't warm to these distributions here, you'll get another chance with hypothesis testing in the next topic.

1 The discrete random variable X has the probability function shown below.

$$P(X = x) = \begin{cases} \dfrac{kx}{6} & \text{for } x = 1, 2, 3 \\[2mm] \dfrac{k(7 - x)}{6} & \text{for } x = 4, 5, 6 \\[2mm] 0 & \text{otherwise} \end{cases}$$

a) Find the value of k.

$k =$ ....
(2 marks)

b) Find $P(1 < X \leq 4)$.

....
(2 marks)

c) Given that x is odd, find $P(3 \leq X < 6)$.

....
(4 marks)

2 The number of points awarded to each contestant in a talent competition is modelled by the discrete random variable X with the following probability distribution:

x	0	1	2	3
$P(X = x)$	0.4	0.3	a	b

A contestant is twice as likely to be awarded 2 points as they are to be awarded 3 points.

By finding the values of a and b, calculate the probability that for two randomly chosen contestants, one scores 2 points and the other scores 3 points.

....
(4 marks)

Statistical Distributions

3 In a game, a player tosses three fair coins. If three heads occur then the player wins 20p.
If two heads occur then the player wins 10p. For any other outcome, the player wins nothing.

a) If X is the random variable 'amount won in pence', draw a table to show the probability distribution of X.

(3 marks)

b) The player pays 10p to play each game. Find the probability that the player wins 40p in total
over two games, given that they make a profit over the two games.

....
(4 marks)

4 5% of chocolate bars made by a particular manufacturer contain a 'golden ticket'.
A student buys 5 of the chocolate bars every week for 8 weeks.

The number of golden tickets he finds is represented by the random variable X.

a) State two necessary conditions for X to follow the binomial distribution B(40, 0.05).

....

....
(2 marks)

Assuming that $X \sim$ B(40, 0.05):

b) Find $P(X > 1)$.

....
(2 marks)

c) Find the probability that more than 35 of the chocolate bars bought by the student
do not contain a golden ticket.

....
(2 marks)

Statistical Distributions

5 A particular model of car, the Dystopia, is prone to developing a rattle in the first year after being made. The probability of any particular Dystopia developing the rattle in its first year is 0.65.

A random sample of 20 one-year-old Dystopias is selected.

a) Find the probability that at least 12 but fewer than 15 of the cars rattle.

....

(3 marks)

b) Find the probability that more than half of the cars rattle.

....

(2 marks)

c) A further five random samples of Dystopias are tested. There are 20 cars in each sample. Find the probability that more than half of the cars in exactly three of these five samples rattle.

Define a new random variable that follows a binomial distribution with $n = 5$ and $p = $ P(more than half of the cars rattle).

....

(3 marks)

6 An ice cream shop owner finds that, on 1st July, 880 out of the 1100 customers chose a sugar cone.

a) A random sample of 20 customers from 2nd July is selected. Use a binomial distribution and the data from the previous day to estimate the probability that exactly 12 of them chose a sugar cone.

Use the info to find p.

....

(4 marks)

The owner claims that 42% of customers buy an ice cream with at least one scoop of chocolate ice cream.

b) Assuming that this claim is correct, and that the next 75 customers form a random sample of customers, find the probability that more than 30 of them choose at least one scoop of chocolate ice cream.

....

(3 marks)

c) Comment on the validity of the binomial model you used in part b).

....

....

(1 mark)

Statistical Distributions

7 The histogram below represents the masses of newborn babies in a particular hospital in 2016. The masses have a mean of 3.55 kg and a standard deviation of 0.5 kg.

a) A nurse models the data using a normal distribution with mean 3.55 kg and standard deviation 0.5 kg. Based on the histogram, do you think this is an appropriate choice of model? Give two reasons to support your answer.

....

....

(2 marks)

b) Babies who have a mass of less than 2.5 kg at birth are said to have a low birth weight. Using the nurse's model, estimate the probability that a randomly-selected newborn baby at the hospital has a low birth weight.

....

(2 marks)

8 The volumes of water in a certain brand of fire extinguisher approximately follow a normal distribution, as shown in the diagram below. The points of inflection of the graph are labelled.

a) Use the diagram to estimate the mean, μ, and standard deviation, σ, for the model.

$\mu =$ .... ml, $\sigma =$ .... ml

(2 marks)

b) Hence use the model to estimate the probability that a randomly-selected fire extinguisher contains between 5900 ml and 6100 ml of water.

....

(2 marks)

Statistical Distributions

9   The exam marks for 1000 candidates can be modelled by a normal distribution with mean 50 marks and standard deviation 15 marks.

     a)   One candidate is selected at random. Find the probability that they scored fewer than 30 marks on this exam.

....

(2 marks)

     b)   The pass mark is 41. Estimate the number of candidates who passed the exam.

....

(2 marks)

     c)   Find, to the nearest whole number, the mark needed for a distinction if the top 10% of the candidates achieved a distinction.

....

(2 marks)

10   The diameters of the pizza bases made at a restaurant are normally distributed. The mean diameter is 12 inches, and 5% of the bases measure more than 13 inches.

     a)   Find the standard deviation of the diameters of the pizza bases to 3 significant figures.

....

(3 marks)

Any pizza base with a diameter of less than 10.8 inches is considered too small and is discarded.

     b)   If 100 pizza bases are made in an evening, approximately how many would you expect to be discarded due to being too small?

....

(3 marks)

Three pizza bases are selected at random.

     c)   Find the probability that at least one of these bases is too small.

....

(3 marks)

Statistical Distributions

11 The time in minutes, X, that it takes a window cleaner to clean each window of an office block is normally distributed with a mean of 4 minutes and a standard deviation of 1.1 minutes.

    a) Find the probability that a randomly-selected window takes less than 3.5 minutes to clean.

        ....
        (2 marks)

    b) Find the probability that the time taken to clean a randomly-selected window deviates from the mean by more than 1 minute.

        ....
        (3 marks)

    c) Find the time taken t, in minutes to 1 decimal place, such that there is a 1% probability that a randomly-selected window will take longer than t minutes to clean.

        .... mins
        (2 marks)

12 The mass, M g, of sweets in a packet is normally distributed, with a mean of 93 g. In 75% of packets of these sweets, there are at most 95 g of sweets.

    a) Packets with less than 88 g of sweets cannot be sold. Calculate the probability that a randomly-selected packet of sweets cannot be sold.

        ....
        (5 marks)

    b) 20 packets are selected at random. Find the probability that exactly one of them cannot be sold.

        ....
        (3 marks)

Statistical Distributions

13 One weekend, 48% of customers at Soutergate Cinema went to see the latest superhero film. A random sample of 200 customers was taken.

a) Write down the binomial distribution that could be used to model S, the number of customers in the sample who went to see the superhero film.

....
(1 mark)

b) Explain why the normal distribution could be used to approximate this binomial distribution.

....

....

....
(2 marks)

c) Specify the normal distribution that could be used to model S, giving the value of any parameters.

....
(2 marks)

The same weekend, 3% of customers at Vulcan Cinema went to see the latest horror film. The manager takes a random sample of 90 customers.

d) Could a normal distribution be used to accurately approximate probabilities for this sample? Explain your answer.

....

....

....
(2 marks)

e) Find the probability that more than 4 of the sampled customers at Vulcan Cinema went to see the horror film.

....
(3 marks)

Make sure you can use the binomial and normal functions on your calculator — you'll need them in the exam. Remember, you can use a normal distribution to approximate a binomial distribution if n is large and p is close to 0.5. You could also be asked to comment on whether a probability model or any assumptions made are appropriate for a given context.

Score

89

Statistical Hypothesis Testing

Here we enter the murky world of hypothesis testing, but we welcome back our friends the binomial and normal distributions. You should remember them from the previous topic... they remember you...

1 Josie wants to find out what pupils in her school think about politics.
   She decides to ask a sample of the pupils.

   a) Identify the population that Josie is interested in.

   ....

   (1 mark)

   b) Josie plans to select everyone in her A-level politics class as her sample. Name Josie's sampling method and explain whether or not her sample is likely to be representative of the population.

   ....

   ....

   (2 marks)

   c) Josie's friend Mike suggests she randomly selects 10 pupils from each year and asks them to send their responses to her questions by email. Suggest one reason why this sampling method may lead to bias.

   ....

   ....

   (1 mark)

2 Jamila is investigating the pay rises given to working adults in her town last year.
   The table below shows the number of working adults in Jamila's town.

Age (in years)	18-27	28-37	38-47	48-57	Over 57
No. of working adults	1200	2100	3500	3200	1500

   Jamila plans to use stratified sampling to select a sample of 50 working adults from her town.

   a) Suggest one reason why it might be sensible for Jamila to stratify her sample by age.

   ....

   ....

   (1 mark)

   b) Jamila wants to investigate the pay rises given to working adults across the UK last year.
   Can she use her sample data to draw conclusions about the whole population? Explain your answer.

   ....

   ....

   (1 mark)

   c) Dareen is also investigating the pay rises given to working adults in the town.
   She samples 50 adults on the town's high street on a Monday afternoon.
   Suggest one reason why Dareen's sample is likely to be biased.

   ....

   ....

   (1 mark)

Statistical Hypothesis Testing

3 For all the Local Authorities in England in 2001, it is known that in 20% of them, fewer than 800 people cycled to work. Adil wants to carry out a hypothesis test to investigate whether the proportion of Local Authorities with fewer than 800 people who cycled to work was different in 2011.

a) Adil uses the large data set to randomly sample 25 English Local Authorities from the 2011 travel data. 4 of the sampled authorities had fewer than 800 people who cycled to work. Use this information to carry out Adil's test at the 10% level of significance.

You can either work out the p-value or find the critical region.

....

....

(5 marks)

b) Adil carries out the same hypothesis test but using a different random sample of 25 authorities and rejects his null hypothesis to conclude that the proportion of authorities with fewer than 800 people who cycled to work was different in 2011. Find the probability that he incorrectly rejected the null hypothesis.

....

(3 marks)

4 Nate teaches judo classes at 'basic' and 'advanced' levels. Last year, 20% of his students had done judo for at least two years. Nate moves to a different judo club and claims that at this club, the percentage of his students who have done judo for at least two years is different. To test this, he surveys a random sample of 20 of his new students.

a) Given that 7 of the sampled students have done judo for at least two years, use a binomial distribution to carry out a test of Nate's claim at the 5% significance level.

....

....

(5 marks)

b) State two assumptions that are needed for the binomial model you used in part a) to be valid. Comment on whether each assumption is likely to be true for Nate's test.

....

....

....

....

(4 marks)

Statistical Hypothesis Testing

5   2016 records suggest that 45% of the members of a gym use the swimming pool. The gym's manager claims that the popularity of the swimming pool has decreased since then. He surveys a random sample of 50 members to test his claim.

After carrying out his test at the 5% level of significance, the manager concludes that there is evidence to suggest that the popularity of the pool has decreased.

Find the maximum possible number of gym members in the sample of 50 who use the pool.

....
(4 marks)

6   A bakery makes different flavours of muffins. The masses of their blueberry muffins can be modelled by a normal distribution with mean 110 g and standard deviation 3 g. A customer claims that the bakery's chocolate muffins weigh less than the blueberry muffins. A random sample of 15 chocolate muffins was found to have a mean mass of 108.5 g.

Assuming that the masses of the chocolate muffins can be modelled by a normal distribution with standard deviation 3 g, test the customer's claims at the 10% level of significance.

....

....
(7 marks)

7   The mean height of the sunflowers in a particular field is 150 cm. The heights of the sunflowers in a second field are known to follow a normal distribution, with a standard deviation of $\sqrt{20}$ cm. The mean height of a random sample of 6 of these sunflowers is 140 cm.

Test at the 1% level of significance whether the average height of the sunflowers in the second field is the same as for the sunflowers in the first field.

....

....
(7 marks)

Statistical Hypothesis Testing

8 The duration in minutes, X, of a car wash is normally distributed with mean 8 and standard deviation $\sqrt{1.2}$. After some maintenance work is carried out, the manager claims that the mean duration of the car wash has decreased. A hypothesis test is to be carried out to investigate this claim.

a) A random sample of 20 washes are timed and the mean duration is found to be 7.8 minutes. Carry out the test at the 5% significance level. You may assume that the standard deviation of the washes' durations is unchanged.

....

....

(7 marks)

b) Find the least value for the sample mean car wash duration that would have provided insufficient evidence to reject the null hypothesis for the test you carried out in part a). Give your answer to 3 significant figures.

....

(2 marks)

9 A government minister reads a report about the number of people who cycle to work in different cities. The report contains the table below for York, made using the UK 2001 census data.

Number of people in employment	Number of people who cycled to work
87 305	10 508

The report also analysed data collected from a random sample of 1000 people, from a city outside the UK in 2001. The number of people in this sample who cycled to work, c, was not included in the report. However, the report did state that there was evidence at the 2.5% significance level to suggest that the proportion of people in employment who cycled to work was lower than in York. Using an appropriate normal distribution, find an estimate for the largest possible value of c.

This question starts as a binomial distribution. A diagram may be useful to help you find the largest possible value of c.

....

(5 marks)

With hypothesis tests, think carefully about whether it's easier to find the p-value or the critical region — and make sure you're an expert at using the binomial functions on your calculator. Plus, one more thing... always write a proper conclusion. For example, don't just say 'reject H_0', you also need to explain what rejecting H_0 means in the context of the question.

Score

56

Correlation and Regression

Correlation and regression are some of my favourite things. In fact, I even named my two dogs 'Correlation' and 'Regression' because I love these topics so much. My children Cora and Reggie thoroughly approved.

1   A construction company measures the length, y metres, of a cable when put under different amounts of tension, T kN (kilonewtons). The results of its tests are shown on the scatter diagram below.

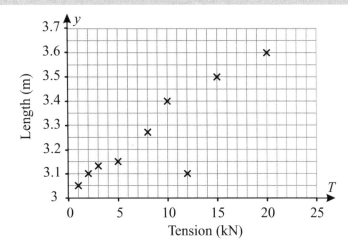

a)   One of the readings has been recorded inaccurately. Put a ring around this reading on the graph.

(1 mark)

b)   Describe the correlation shown on the scatter diagram.

....

(1 mark)

c)   Use the scatter diagram to complete the table below.

T (kN)	1	2		5	8	10	12		20
y (metres)		3.1	3.13	3.15	3.27		3.1	3.5	3.6

(2 marks)

An engineer believes a linear regression line of the form $y = a + bT$ could be used to accurately describe the results. She ignores the outlier, and calculates the equation of the regression line to be $y = 3 + 0.03T$.

d)   Explain what these values of a and b represent in this context.

....

....

(2 marks)

The engineer wants to use the regression line to predict the length of the cable when it is put under a tension of 50 kilonewtons.

e)   Comment on the reliability of her estimate.

....

....

(1 mark)

Correlation and Regression

2 For Local Authorities in Yorkshire and The Humber in 2011, the scatter diagram below shows the total population size and the percentage of people aged between 16 and 74 not in employment.

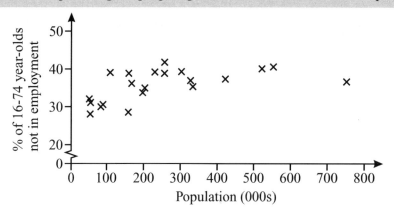

a) Describe the type of correlation shown on the scatter diagram.

....

(1 mark)

b) Tom suggests that Local Authorities with larger populations generally have higher rates of unemployment, although he admits that there are some exceptions.
Circle any data points on the graph above that Tom might consider to be exceptions.

(1 mark)

3 Some ecologists carry out an investigation into the reindeer populations at different locations.
They also recorded the human population density at the same locations.
Their results are displayed in the scatter diagram below.

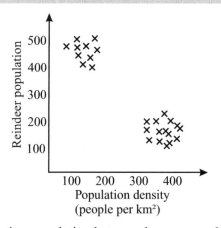

a) Jiao claims that there is negative correlation between human population density and reindeer population.
Use the scatter diagram to comment on Jiao's claim.

....

....

(2 marks)

b) Killian claims that more people living in an area cause there to be fewer reindeer in that area.
Do you agree with Killian's claim? Explain your answer.

....

....

(2 marks)

Correlation and Regression

4 A student collects 10 pairs of data values from an estate agent's website.
Each pair of values shows the floor area (A) and the advertised price (P) of a randomly-selected house.

Using her 10 pairs of values, the student calculates the PMCC between A and P to be 0.634.
Carry out a hypothesis test at the 5% significance level of whether the PMCC between A and P for the
population of houses advertised on the whole website is positive. State your hypotheses clearly.

Sample	PMCC Significance Level				
Size	0.10	0.05	0.025	0.01	0.005
10	0.443	0.549	0.632	0.716	0.765

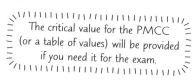
The critical value for the PMCC (or a table of values) will be provided if you need it for the exam.

....

....

(4 marks)

5 The scatter diagrams below show the population size and annual CO_2
emissions of the 10 largest cities in two different countries, A and B.

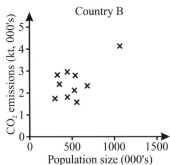

a) Dom says that the outlier in the scatter diagram for Country B should be excluded
from all analysis of the data. Explain whether or not Dom's suggestion is reasonable.

....

....

(1 mark)

The product moment correlation coefficient for the 10 cities in Country A is 0.833.
The critical value for a 2-tail test at the 1% significance level is 0.765.

b) Carry out a hypothesis test at the 1% significance level of whether the PMCC for this data is non-zero.
State your hypotheses clearly.

....

....

(3 marks)

EXAM TIP Scatter diagrams, regression lines and the PMCC all help you to understand relationships
between different sets of data — but just because two data sets are correlated, that doesn't
necessarily mean that one causes changes in the other. Apparently, ice cream sales and murder
rates are correlated... but that doesn't mean the next person you see eating ice cream is a killer.

Score

21

Kinematics 1

Kinematics is all about the way things move — by the end of this topic your flamboyant dance skills should be most impressive. For questions in this topic give your answers to an appropriate degree of accuracy.

1 A particle moves in a straight line with a constant acceleration. The initial velocity of the particle is U ms^{-1} and its velocity at time T is V ms^{-1}, as shown on the velocity-time graph below.

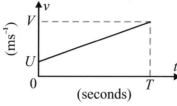

Using the graph, show that the displacement of the particle can be given by $S = \frac{1}{2}(U + V)T$.

....

(3 marks)

2 A ball is thrown vertically upwards with velocity 5 ms^{-1} from a point 2 m above the ground. The velocity-time graph below models the motion of the ball. Take g = 9.8 ms^{-2}.

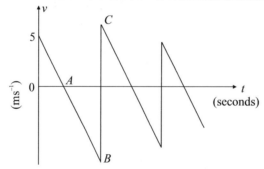

a) Find the time taken by the ball to reach point A.

....

(3 marks)

b) Find the velocity of the ball when it reaches point B.

....

(3 marks)

c) Explain why, in reality, this graph may not be an accurate model of the velocity of the ball.

....

(1 mark)

Kinematics 1

3 A train is moving along a straight horizontal track with constant acceleration.
The train enters a tunnel which is 760 m long and exits 24 seconds later.
8 seconds before entering the tunnel, the train passes a sign. The sign is 110 m from the tunnel's entrance.

110 m   760 m

Find the train's acceleration and its speed when it passes the sign.

$a =$ .... $u =$ ....

(7 marks)

4 A robot initially has position vector $(2\mathbf{i} + \mathbf{j})$ m and is travelling with constant velocity $(\mathbf{i} + 3\mathbf{j})$ ms⁻¹.
After 8 seconds it reaches point A. A second robot has constant velocity $(-5\mathbf{i} + 2\mathbf{j})$ ms⁻¹
and takes 5 seconds to travel from point A to point B.

a) Calculate the difference in speed between the two robots.

....

(2 marks)

b) Find the direction of motion of the second robot, measured anticlockwise from
the positive $\mathbf{i}$ direction.

....

(2 marks)

c) Find the position vectors of points A and B.

$A =$ ....

$B =$ ....

(2 marks)

Kinematics 1

5 At time $t = 0$, a particle, P, sets off from rest at the point O and accelerates uniformly. At time $t = 3$ seconds, the particle has position vector $(9\mathbf{i} - 18\mathbf{j})$ m relative to O.

a) Given that the particle continues to move with this constant acceleration, find an expression for the position vector, $\mathbf{p}$, of the particle at time t.

$\mathbf{p} = $ ....

(4 marks)

Also at time $t = 0$, a second particle, Q, moving with constant velocity $(3\mathbf{i} - 5\mathbf{j})$ ms^{-1} has position vector $(a\mathbf{i} + b\mathbf{j})$ m, where a and b are constants.

b) Given that the two particles collide at time $t = 8$ seconds, find the values of a and b.

$a = $ .... $b = $ ....

(3 marks)

6 A ball's velocity is modelled by the vector $\mathbf{v} = [(2t - 5)\mathbf{i} + (3t - 8)\mathbf{j}]$ ms^{-1}, where $\mathbf{i}$ and $\mathbf{j}$ are unit vectors pointing east and north respectively. At time $t = 2$ s, the ball is at position vector $(-2\mathbf{i} + 10\mathbf{j})$ m.

a) Find the value of t at which the ball is travelling directly eastwards.

$t = $ ....

(2 marks)

b) Calculate the range of t for which the ball is to the west of the origin.

....

(5 marks)

Kinematics 1

7 A particle P sets off from the origin at $t = 0$ and starts to move along the x-axis in the direction of increasing x. After t seconds, P has velocity v ms^{-1}, where v is given by:

$$v = \begin{cases} 11t - 2t^2 & 0 \leq t \leq 5 \\ 25 - 4t & t > 5 \end{cases}$$

a) Find the displacement of P from the origin at $t = 5$.

....

(4 marks)

b) Find the time taken for P to return to the origin.

....

(5 marks)

8 An object has constant acceleration a and an initial velocity of u.

Given that the object sets off from the origin, use integration to show that the object's displacement at time t is given by the formula $s = ut + \frac{1}{2}at^2$.

(4 marks)

Kinematics 1

9 A particle moves in a straight line, beginning at the origin at $t = 0$.
Its velocity, v ms^{-1}, at time t seconds is given by $v = 2t - 3e^{-2t} + 4$.

a) Find an expression for the particle's acceleration, a ms^{-2}, at time t.

$a = $ ....

(2 marks)

b) Find the range of values for the particle's acceleration.

....

(3 marks)

c) Find an expression for the particle's displacement from O, s m, at time t.

$s = $ ....

(3 marks)

10 A particle is moving in a curved path. Its velocity, **v** ms^{-1}, at time t seconds is given by
v $= (2\cos 3t + 5t)$**i** $+ (2t - 7)$**j**, where the unit vectors **i** and **j** are in the directions of east and north respectively.

a) Find an expression for the particle's acceleration, **a** ms^{-2}, after t seconds.

a $= $ ....

(2 marks)

b) Calculate the magnitude of the particle's maximum acceleration. Give an exact value.

....

(3 marks)

Kinematics questions can be tough, because they combine lots of skills — you need to be comfortable with using vectors, differentiating and integrating, as well as all the mechanics-y stuff. Just remember, whenever you see 'constant' or 'uniform' acceleration in an exam question you should think 'suvat'. If acceleration isn't constant, then it's time to whip out the calculus.

Score

63

Kinematics 2

These questions are all about projectiles. Master the maths of trajectories and you'll be a pro-golfer in no time. Give your answers to an appropriate degree of accuracy and take $g = 9.8$ ms^{-2}, unless told otherwise. Fore!

1 A golf ball is struck from a point T on horizontal ground with velocity $(29\mathbf{i} + p\mathbf{j})$ ms^{-1}, at an angle α to the horizontal. The ball takes 5 seconds to land on the ground at point G.

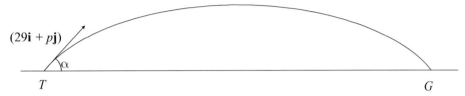

a) Find p.

$p = $ ....
(2 marks)

b) Find the horizontal distance TG.

....
(2 marks)

2 A stone is thrown upwards from point A on a cliff, 22 m above horizontal ground, with speed 14 ms^{-1} at an angle of 46° to the horizontal. After projection, the stone moves freely under gravity and lands at B on the horizontal ground, as shown below.

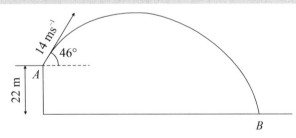

a) Find the length of time for which the stone is at least 22 m above the ground.

....
(4 marks)

b) Calculate the speed of the stone as it hits the ground.

....
(5 marks)

Kinematics 2

3 A stone is thrown from A, at a height of 1 m above the ground. The stone's initial velocity is 10 ms^{-1} at an angle of 20° above the horizontal. The stone lands on the horizontal ground at B.

a) Calculate the maximum height above the horizontal ground that the stone reaches during its flight.

....

(4 marks)

b) Find the time it takes the stone to travel from A to B.

....

(3 marks)

c) Explain why the answer you found in part b) might not be the actual time it takes the stone to travel from A to B.

....

....

(1 mark)

Projectiles questions can look really complicated. The trick is: take a deep breath, write out the variables you've been given in the question and then think which suvat equation will let you work out the next bit of the question. Sometimes, you'll have to go through this process a couple of times before you get to the final answer, but so long as you're careful you'll get there.

Score

21

Forces and Newton's Laws

Try as you might, there's no escaping Newton's laws — they pretty much govern the way everything moves. In this topic, you should give your answers to an appropriate degree of accuracy and take $g = 9.8$ ms^{-2}.

1   Three forces act at a point, O, as shown below.

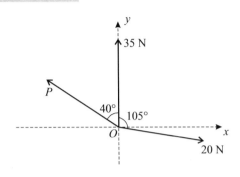

a)   Given that there is no resultant force in the direction of the x-axis, find P.

$P =$ ....
(2 marks)

b)   Find the magnitude and direction of the overall resultant force, R.

....

(3 marks)

2   Two forces, $\mathbf{A} = (2\mathbf{i} - 11\mathbf{j})$ N and $\mathbf{B} = (7\mathbf{i} + 5\mathbf{j})$ N, act on a toy helicopter of mass 500 g.

a)   Find the magnitude of the resultant force.

....
(3 marks)

b)   Find the acceleration of the helicopter, $\mathbf{a}$ ms^{-2}.

$\mathbf{a} =$ ....
(2 marks)

Forces and Newton's Laws

3   A 2 kg ring is threaded on a rough horizontal rod. A rope is attached to the ring, and is held at 40° below the horizontal, as shown. The normal contact force of the rod on the ring is R N, the tension in the rope is S N and frictional force between the rod and the ring is F N.

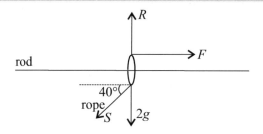

The coefficient of friction between the rod and the ring is 0.3.
Given that the ring is stationary, find the range of possible values that S can take.

$S =$ ....

(5 marks)

4   A horizontal force of 25 N causes a particle of mass 7 kg to accelerate up a rough plane inclined at 15° to the horizontal, with acceleration of magnitude 0.2 ms^{-2}, as shown on the diagram to the right.

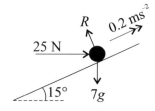

a)   Calculate the coefficient of friction between the mass and the plane.

$\mu =$ ....

(6 marks)

b)   The 25 N force is now removed and the particle is released from rest. Find how long the particle takes to slide a distance of 3 m down the plane.

....

(6 marks)

Forces and Newton's Laws

5 A block of mass m kg rests on rough horizontal ground. The block is held in limiting equilibrium by a string under tension T N pulling upwards at an angle of 51.3° to the horizontal. The coefficient of friction between the block and the ground is 0.6 and the magnitude of the frictional force is 1.5 N.

a) Sketch a diagram showing the block and the forces acting upon it.

(1 mark)

b) Find the tension in the string, T, and the mass of the block, m.

$T =$ ....

$m =$ ....

(5 marks)

6 A ball of weight W N is attached to a string under tension T N on a rough plane inclined at β to the horizontal. The string is parallel to the plane. The ball is held in limiting equilibrium by a frictional contact force F N, acting in the same direction as T.

Assuming that T acts parallel to the plane, show that the coefficient of friction $\mu = \tan \beta - \dfrac{T}{W}\sec \beta$.

(5 marks)

Forces and Newton's Laws

7 A woman is travelling in a lift. The lift is rising vertically and is accelerating at a rate of 0.75 ms^{-2}. The lift is pulled upwards by a light, inextensible cable. The tension in the cable is T N and the lift has mass 500 kg. The floor of the lift exerts a force of 675 N on the woman. Find T.

$T =$ ....

(4 marks)

8 One end of a light inextensible string is attached to a block A, of mass 35 kg. The string passes over a fixed, smooth pulley and is attached at the other end to a block B, of mass M kg.

The system is held at rest by a vertical light inextensible string attached to B at one end and to horizontal ground at the other, as shown. The tension in this second string is K N.

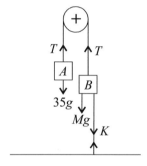

a) Find an expression for K in terms of M and g.

....

(2 marks)

b) (i) The string fixing B to the ground is cut. Three seconds after the string is cut, A is travelling at 1 ms^{-1} towards the ground. Calculate M.

$M =$ ....

(5 marks)

(ii) Give one assumption you have made about the behaviour of the system after the string is cut.

....

(1 mark)

Forces and Newton's Laws

9 A particle, A, is attached to a weight, W, by a light inextensible string which passes over a smooth pulley, P, as shown. When the system is released from rest, with the string taut, A and W experience an acceleration of 4 ms^{-2}. A moves across a rough horizontal plane and W falls vertically. The mass of W is 1.5 times the mass of A.

a) Given that the mass of A is 0.2 kg, find the coefficient of friction, μ, between A and the horizontal plane.

$\mu = $ ....

(6 marks)

W falls for h m until it impacts the ground and does not rebound. A continues to move until it reaches P with speed 3 ms^{-1}. The initial distance between A and P is $\frac{7}{4}h$.

b) Find the time taken for W to impact the ground.

....

(7 marks)

c) How did you use the information that the pulley is smooth?

....

....

(1 mark)

Forces and Newton's Laws

10 A particle A of mass 3 kg is placed on a rough plane inclined at an angle of $\theta = \tan^{-1}\frac{3}{4}$ above the horizontal and is attached by a light inextensible string to a second particle B of mass 4 kg.

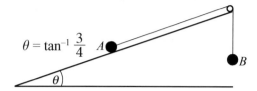

The string passes over a smooth pulley at the top of the inclined plane so that particle B hangs freely. When the system is released, particle B moves downwards vertically with an acceleration of 1.4 ms^{-2} and A moves up the plane.

a) Find the coefficient of friction between particle A and the plane.

$\mu = $ ....

(7 marks)

Two seconds after the particles are released, the string breaks. A then continues up the plane until it comes instantaneously to rest. At no point does A reach the pulley.

b) Find how far particle B moves from the start of the motion until the string breaks.

....

(2 marks)

c) Calculate how far particle A moves from the instant the string breaks until it comes to rest.

....

(5 marks)

Moments

It had to happen sooner or later, and there's no putting it off any longer — it's your favourite topic... moments. As ever, give your answers to an appropriate degree of accuracy and take $g = 9.8 \text{ ms}^{-2}$, unless told otherwise.

1 A uniform rod AB of length 4 m and weight 20 N hangs in equilibrium in a horizontal position supported by two vertical inextensible strings attached at A and B. A bead of weight 16 N rests on the rod at a point C, which is 3 m from A. The bead can be modelled as a particle.

a) Find the tensions in the strings at A (T_1) and at B (T_2).

$T_1 = $ ....

$T_2 = $ ....

(3 marks)

b) The bead is now moved along the rod so that the tension in the string at B is twice the tension in the string at A. Find the distance of the bead from A.

....

(3 marks)

2 A skateboarder balances a skateboard on a rail. The skateboard has mass 4 kg and can be modelled as a uniform horizontal rod, AB, 0.8 m in length. The skateboard is in contact with the rail at point C, where $AC = 0.5$ m. The skateboarder has mass 80 kg and places her feet at points X and Y. The skateboard is in equilibrium when the force applied at Y is three times the force applied at X and CY is half the magnitude of CB.

a) Find the distance AX.

....

(5 marks)

b) Find the magnitude of the normal contact force, R, at the rail.

....

(2 marks)

c) An object of mass 3 kg is placed on the skateboard at point A. The skateboarder moves her foot such that AX is increased by 13 cm. The skateboard remains horizontal and in equilibrium. Find the magnitude of the forces applied at X (F_X) and Y (F_Y).

$F_X = $ ....

$F_Y = $ ....

(5 marks)

Section Three — Mechanics

Moments

3 A freely-hinged beam, *AB*, attached to a vertical wall is held in equilibrium perpendicular to the wall by a strut attached to the beam at *C*. The strut is fastened to the wall at point *D*, making an angle of 60° with the wall, as shown. A particle, *P*, with a mass of 1.8 kg rests upon the beam at a point 0.2 m from *A*. The weight of the beam acts at point *C*. The distance from *A* to the centre of mass of the beam is 1.1 m.

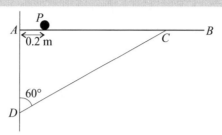

a) Find the magnitude of the vertical component of the reaction at *A*.

....

(3 marks)

The magnitude of the horizontal component of the reaction at *A* is 35 N.

b) Find the magnitude of the reaction at *A* and the angle it makes with the horizontal.

....

(3 marks)

c) Calculate the tension in the strut.

....

(3 marks)

4 A shape is made up of a uniform square lamina and a uniform rectangular lamina as shown on the right. The centre of mass of the shape is located 2 m from AB and 2.28 m from AC.

The shape is smoothly pivoted at A and rests in equilibrium against a smooth vertical wall. The angle between AC and the horizontal ground is 30°. Given that the weight of the shape is 4 N, find the magnitude of the reaction force, *R*, of the wall on the shape.

Start by finding the components of the forces parallel and perpendicular to AC.

R = ...

(4 marks)

Moments

5 A uniform rod, *AB*, rests in limiting equilibrium against rough horizontal ground and is held at an angle of θ to the horizontal by a smooth peg, *C*, where $\cos\theta = 0.91$. *AB* is 3 m in length and has a mass of 8 kg. The distance $CB = l$ m. The magnitude of the normal reaction at the peg is 54 N.

a) Find *l*.

$l = $ ....

(3 marks)

b) Find μ, the coefficient of friction between the ground and the rod.

$\mu = $ ....

(6 marks)

c) A vertical force of magnitude *V* is applied at B and just begins to make the rod rotate around the peg C. Show that *V* is given by $V = 4g\left(\frac{3}{l} - 2\right)$.

(2 marks)

6 A uniform ladder has mass 10 kg and length 8 m. It leans against a rough vertical wall, making an angle of θ with the rough horizontal ground. A painter of mass 62 kg stands on the ladder three-quarters of the distance between the bottom of the ladder and the top. The coefficient of friction between the ladder and the wall is 0.5 and the coefficient of friction between the ladder and the ground is 0.4.
If the ladder is in limiting equilibrium, show that $\tan\theta = \frac{79}{48}$.

(7 marks)

108

Moments

7   A uniform ladder of length 8 m and weight 30 N has one end resting against a smooth vertical wall at point B and the other end on rough horizontal ground at point A, as shown below. The angle between the ladder and the ground is 70° and the coefficient of friction (μ) between the ladder and the ground is 0.1.

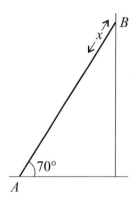

A firefighter with weight 805 N stands on the ladder a distance x from B. The firefighter's colleague pushes the bottom of the ladder horizontally with her foot in the direction of the wall to prevent it sliding. The reaction force from the wall is 260 N.

a)   Find the distance x, given that the ladder is in equilibrium.

$x =$ .... m

(3 marks)

b)   Find the minimum horizontal force, P N, required by the colleague's foot to prevent the ladder from sliding.

$P =$ ....

(5 marks)

c)   Explain how you've used the assumption that the ladder is uniform.

....

(1 mark)

EXAM TIP

If I had to sum up this topic in 5 words, I'd say this: "Equate moments and balance forces." Okay, fair enough, you have to be able to actually do the equating of moments and balancing of forces, and if I had about 800 more words to play with, I'd talk you through that. But used wisely I reckon those 5 words will get you around 80% of the marks on these pages.

Score

58

Section Three — Mechanics

General Certificate of Education
Advanced Level

A-Level Mathematics
Practice Exam Paper 1: Pure Mathematics

Time Allowed: 2 hours

There are 100 marks available for this paper.

Formulas are given on pages 164 to 166.

1 $$f(x) = x^3 - 2x^2 - 13x - 10$$

    a)   Show that $(x + 1)$ is a factor of f(x).

                                                      (2 marks)

    b)   Hence, or otherwise, solve f(x) = 0.

                                                      (3 marks)

2   Write $2 \log a^3 b - \log ab$ as a single logarithm in its simplest form.

                                                      (3 marks)

3   The functions f, g and h are defined below.

    $f(x) = \dfrac{2x + 7}{3x - 5}, x \in \mathbb{R}, x \neq \dfrac{5}{3}$ $g(x) = x^2 - k, x \in \mathbb{R}$ $h(x) = \cos x, x \in \mathbb{R}$
                                                            where k is a positive constant

    a)   Find $f^{-1}(x)$.

                                                         (2 marks)

    b)   gf(2) = 120. Find the value of k.

                                                         (3 marks)

    c)   Find gh(x) and hg(x), and hence show that the range of gh(x) is not the same as the range of hg(x).

                                                         (4 marks)

4   For small values of θ, show that $\dfrac{1 - \tan 3\theta}{\cos 4\theta - \sin \theta^2} \approx \dfrac{1}{1 + 3\theta}$.

                                                         (3 marks)

5 A sequence is defined by the recurrence relation $u_{n+1} = 5u_n - 2$, where $n \geq 1$.

   a) Given that $u_3 = 38$, find u_1.

(2 marks)

   b) Suppose instead that the sequence is decreasing. Find the range of possible values of u_1.

(2 marks)

   c) Find a value of u_1 for which $\sum_{n=1}^{r} u_n = \frac{1}{2}r$ for any value of r.

(1 mark)

6 Prove that $\sqrt{2}$ is an irrational number.

(4 marks)

7 **a)** Express $\dfrac{11x - 7}{(2x - 4)(x + 1)}$ in partial fractions.

(4 marks)

   b) Hence find the binomial expansion of $\dfrac{11x - 7}{(2x - 4)(x + 1)}$, in ascending powers of x, up to and including the term in x^2.

(6 marks)

8 The curve C has parametric equations

$$x = 1 + 5 \cos t, \quad y = 2 + 5 \sin t, \quad 0 \leq t \leq 2\pi$$

   a) Show that the Cartesian equation of the curve C is the circle given by the equation

$$(x - 1)^2 + (y - 2)^2 = 25$$

(3 marks)

The line l is the tangent to the circle at the point P (4, 6).

   b) Find the equation of the line l. Give your answer in the form $ax + by + c = 0$, where a, b and c are integers to be found.

(4 marks)

9 **a)** Find the coordinates of the stationary point on the curve $y = x \ln x$.
        Give your answer in exact form.

(5 marks)

   b) Determine the nature of the stationary point. Give a reason for your answer.

(2 marks)

10 In trapezium $ABCD$, the acute angle $BAD = x$ radians. Angles BCD and ABC are $\frac{\pi}{2}$.
$DA = 5$ cm and $AB = 2CD$, as shown in the diagram.

a) Show that an expression for the perimeter, P cm, can be written as $P = R \sin (x + \alpha) + 5$,
where R and α are values to be found.

(7 marks)

b) Given that the perimeter of the trapezium is 17 cm, find the value of x to 3 significant figures.

(3 marks)

11 The rate of growth of a population of bacteria is proportional to the number of bacteria in the population.

a) Show that the number of bacteria in the population, P, can be modelled by the function $P = Qe^{kt}$,
where t is the time in hours after the population was first observed and Q and k are constants.

(3 marks)

b) When the experiment begins, there are 5300 bacteria. After 6 hours, there are 876 bacteria present.
Find k to 1 decimal place.

(3 marks)

c) Use your value of k to describe how the population of bacteria is changing.

(1 mark)

12 The parallelogram $ABCD$ has sides of length 5 units and 9 units.

a) The coordinates of A are $(2, 4, 1)$. B has coordinates $(9, 8, z)$.
Find the two possible values of z.

(4 marks)

b) Given that z is a positive integer and that the unit vector in the direction of
$\overrightarrow{BC}$ is $0.8\mathbf{j} + 0.6\mathbf{k}$, show that the coordinates of C are $(9, 12, 8)$.

(3 marks)

c) Hence find the exact length of the diagonal $\overrightarrow{AC}$.

(3 marks)

13 **a)** Find $\int \frac{\ln x}{x^5} \, dx$.

(4 marks)

b) **In this question you must show detailed reasoning**

Find $\int_5^{10} \frac{2x}{\sqrt{x-1}} \, dx$, using the substitution $u = \sqrt{x-1}$.

(5 marks)

14 Part of a concrete sculpture is modelled as a cube with sides of length x m. Concrete costs £65 per cubic metre. Each face is then covered in gold leaf which costs £150 per square metre.

a) Given that the cost of materials used for the cube is £250, prove that x satisfies the equation
$$x = \sqrt{\frac{250 - 65x^3}{900}}$$

(4 marks)

b) Hence show that the length of the cube is between 0.5 m and 1 m.

(2 marks)

c) Using the iteration formula $x_{n+1} = \sqrt{\frac{250 - 65x_n^3}{900}}$ with $x_1 = 0.75$,

find x_2 and x_5. Give your answers correct to 4 decimal places.

(2 marks)

d) Use your value for x_5 to estimate the percentage of the total cost of the cube that was spent on concrete. Give your answer correct to 1 decimal place.

(2 marks)

The graph below shows the curve $y = \sqrt{\frac{250 - 65x^3}{900}}$ and the line $y = x$. The position of x_1 is indicated.

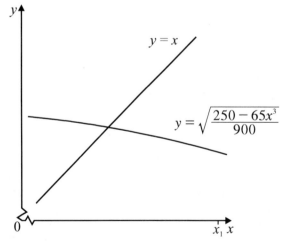

e) Use the graph to show the positions of x_2 and x_3.

(1 mark)

END OF EXAM PAPER

TOTAL FOR PAPER: 100 MARKS

General Certificate of Education
Advanced Level

A-Level Mathematics
Practice Exam Paper 2: Pure Mathematics and Statistics

Time Allowed: 2 hours

There are 100 marks available for this paper.

Formulas are given on pages 164 to 166.

Section A: Pure Mathematics

1 a) $f(x) = \dfrac{x^3 - 9x^2 + 14x}{x^2 - 4} \times \dfrac{x + 2}{x}$. Write $f(x)$ in its simplest form.

(3 marks)

b) Find the coordinates of the points of intersection between $y = |f(x)|$ and the line $y = -\dfrac{1}{2}x + 5$.

(4 marks)

2 A curve has parametric equations:

$$x = t - 3, \; y = (t + k)^2 \text{, where } k \text{ is a constant.}$$

Given that the curve passes through the point $(-7, 0)$, find the value of k.

(3 marks)

3 The diagram below shows a rectangular field 20 m long and 8 m wide enclosed by fences.
Inside the field is a rabbit pen, which is a regular hexagon with sides of length 2 m.
The pen is positioned with one of its sides up against one of the 20 m fences.

A dog is tied to the point labelled D on the diagram. The dog's lead is 2 m long. It is not able to enter the pen.

a) Calculate the exact area of the field the dog can reach.

(3 marks)

b) A cat now enters the field. It stays out of the area that the dog can get to and it is
not able to enter the rabbit pen. Find the area of the field that the cat might go in.
Give your answer correct to 3 significant figures.

(4 marks)

4   A buoy floats on the surface of the sea.
The height, H metres above the sea bed, of the buoy at time t hours is modelled by the equation

$$H = 2\cos\left(\frac{4\pi t}{25}\right) + 4, \quad 0 < t \le 12$$

a)   Find $\dfrac{dH}{dt}$.

(1 mark)

b)   Find the rate of change of the height of the buoy after 7 hours.

(2 marks)

c)   Find the minimum height of the buoy above the sea bed.

(2 marks)

5   **In this question you must show detailed reasoning.**

Solve the equation $2\csc^2 x + 5\cot x = 9$, for $-90° < x < 90°$.
Give your answers to 1 decimal place where appropriate.

(6 marks)

6   **a)**   Find $f'(x)$ if $f(x) = 4x^3 + e^{x^2}$.

(3 marks)

b)   Show that the graph of $f(x)$ is convex when $x > 0$.

(5 marks)

7   The diagram on the right shows the graph of $y = \dfrac{e^{\sin x} + x}{(x+1)^3}$ for $0 \le x \le 2$.

The table below shows values of y for values of x between 0 and 2.
The y-values have been rounded to 3 and 5 decimal places.

x	0	0.5	1	1.5
y (3 d.p.)	1		0.415	0.270
y (5 d.p.)	1		0.41497	0.26953

a)   Complete the table.

(1 mark)

b)   Using the trapezium rule with 3 strips, and the values of y that have been rounded to 3 decimal places, find an approximation for $\displaystyle\int_0^{1.5} \frac{e^{\sin x} + x}{(x+1)^3}\,dx$.

(3 marks)

c)   Sophie calculates that the actual area bounded by the curve, the x-axis, the y-axis and the line $x = 1.5$ is 0.8224, correct to four decimal places. She says that the estimate found using the trapezium rule was larger than this because each rounded value of y was larger than its true value.
By considering the graph above, comment on Sophie's claim.

(1 mark)

8   A pan containing soup at a temperature of 95°C is removed from a stove and the soup is left to cool in a kitchen. t minutes after being removed from the stove, the temperature of the soup is T °C.

The rate of heat loss of the soup over time is modelled as being directly proportional to the difference in temperature between the soup and the kitchen.
The temperature of the kitchen is 15°C.

a)   Explain why $\frac{dT}{dt} = -k(T - 15)$, for some positive constant k.

(1 mark)

b)   Find an expression for T in terms of t and k.

(3 marks)

After 10 minutes, the temperature of the soup is 55°C.

c)   Find the exact value of k, and hence show that $T = 80(0.5)^{pt} + 15$
where p is a positive constant to be found.

(4 marks)

d)   State one assumption that has been made for this model.

(1 mark)

END OF SECTION A

Section B: Statistics

9   Elena is investigating the ages of residents in County Durham.
She uses the 2011 data from the large data set to calculate the following summary statistics, where x is the midpoint of the age classes (in years):
$$\sum f = 513\ 242, \quad \sum fx = 21\ 064\ 931, \quad \sum fx^2 = 1\ 141\ 982\ 145$$

a)   Use this information to estimate the mean and standard deviation of the ages of County Durham residents in 2011.

(3 marks)

Elena believes that the data for County Durham contains outliers at both ends of the data.

b)   Estimate the limits beyond which any outliers in the 2011 data would lie, and comment on Elena's belief.

(2 marks)

Elena lives in a village in County Durham. The oldest person currently living in her village is 87 years old. Elena therefore claims that none of the upper outliers from the 2011 data were living in her village.

c)   Explain why Elena may be wrong in her claim.

(1 mark)

d)   Elena estimates the mean and standard deviation for the ages of residents in Newcastle upon Tyne using the 2011 data from the large data set. She finds that the mean is 37.1 years and the standard deviation is 22.5 years (both to 3 s.f.). Compare the ages of the residents of County Durham and Newcastle upon Tyne in 2011.

(2 marks)

10   Chloe visits her local post office each weekday. The probability that she has to queue
for longer than 2 minutes on any visit is 0.4.

a)   She records how long she has to wait on a random sample of 10 visits to the post office.
She decides to model this information using a binomial distribution.
State two assumptions Chloe has made for this model.

(2 marks)

b)   Chloe then records how long she has to wait on a random sample of 50 visits to the post office.
She decides to use a normal approximation to the binomial distribution to model this information.
Find the normal approximation she should use, stating any parameters clearly, and explain why a
normal approximation is appropriate in this situation.

(2 marks)

Chloe also wants to survey some of the residents of her town to ask their opinion about the
service provided by the post office. She obtains a copy of the electoral roll and selects one person
at random from the first 20 people listed. She then surveys every 20th person on the electoral roll.

c)   What type of sampling has Chloe used to collect her data?

(1 mark)

11   The scatter diagrams below show the percentages of the population aged 18-29 and the percentages
of the workforce who walk to work within each Local Authority in two different regions in 2011.

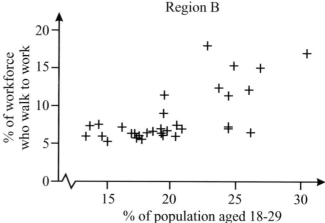

a)   Describe the correlation shown on the scatter diagram for Region A,
and explain what it means in context.

(1 mark)

b)   One scatter diagram shows information for London and the other shows information for the
West Midlands. Match each scatter diagram to each region. You must justify your answer.

(2 marks)

c)   One Local Authority in Region B was excluded from the scatter diagram. For this Local Authority,
21.6% of the population were aged 18-29, and 45.8% of them walked to work. Do you think it was
reasonable to exclude this value from the diagram? Explain your answer.

(1 mark)

12 Hashim wants to investigate the age distribution of people aged 20-90 in Wales in 2001.
He represents the information from the large data set on a histogram, shown below.

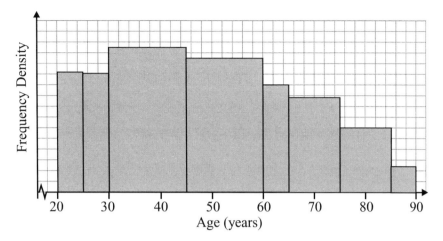

a) Hashim suggests using the histogram to calculate the mean (μ) and standard deviation (σ), then modelling the data using a normal distribution $N(\mu, \sigma^2)$.
With reference to the histogram, explain why this would not be an appropriate model.

(1 mark)

b) Leanne collects her own age data from a sample of adults in her town in Wales.
For her data, she calculates the mean and standard deviation to be 54 years and 12.5 years respectively. The values for the upper and lower quartiles are $Q_3 = 62$ years and $Q_1 = 45$ years.
Based on these statistics, explain whether the normal distribution $N(54, 12.5^2)$ could be a suitable model for Leanne's data. Use calculations to support your answer.

(2 marks)

13 A juice manufacturer produces orange juice in cartons. The volume of juice in a carton, J ml, is assumed to be normally distributed. Based on previous data, the quality control officer finds that 5% of cartons contain less than 493 ml, and 2.5% contain more than 502 ml.

a) Find the mean and standard deviation of J.

(5 marks)

A carton meets the manufacturer's standards if it contains more than 492 ml of juice.

b) Find the probability of a carton failing to meet the manufacturer's standards.

(2 marks)

In an effort to cut costs, the manufacturer modifies the production process. After this process has been in place for a month, a random sample of 20 cartons are found to contain a mean volume of 498.7 ml of juice.

c) Assuming the standard deviation is unchanged, test, at a 1% level of significance, whether or not the mean volume of juice in a carton has changed under the new production process.
You must state your hypotheses clearly.

(7 marks)

14 Adam usually goes to the gym on Saturdays. When he does, he uses only the swimming pool or the rowing machines. On any given Saturday:

R represents the event that he uses the rowing machines.

S represents the event that he uses the swimming pool.

$$P(R \cup S) = \frac{8}{10} \quad P(R|S) = \frac{4}{9} \quad P(S|R) = \frac{4}{11}$$

a) Calculate P(R) and P(S).

(5 marks)

b) Explain whether the events R and S are independent.

(2 marks)

T represents the event that Adam has a takeaway on any given Saturday.

The events S and T are independent.

$$P(T) = 0.15 \quad P(R \cap T) = 0.08 \quad P(R \cap S \cap T) = 0.05$$

c) Draw a Venn diagram to show events R, S and T, giving the probabilities for each region.

(4 marks)

15 Alice and Ben are studying the approximate number of a particular bacteria, y, in ponds of different maximum depths, x m. They collect data from 8 ponds, with maximum depths varying from 0.5 m to 12 m, and calculate the product moment correlation coefficient between x and y for their data. Alice obtains a value of $r_A = 0.71$, and Ben calculates his value to be $r_B = 1.09$.

a) Interpret, in context, the value of Alice's correlation coefficient, and explain how you can tell that Ben must have made a mistake in his calculation.

(2 marks)

b) Alice claims that her results show that deeper ponds result in higher numbers of bacteria. Comment on this claim.

(1 mark)

c) The critical value for a 1-tail test at a 5% significance level on a sample size of 8 is 0.6215. Stating your hypotheses clearly, test at the 5% significance level whether Alice's results are statistically significant evidence of positive correlation between the number of this type of bacteria and pond depth in the population of ponds.

(2 marks)

END OF EXAM PAPER

TOTAL FOR PAPER: 100 MARKS

General Certificate of Education
Advanced Level

A-Level Mathematics
Practice Exam Paper 3: Pure Mathematics and Mechanics

Time Allowed: 2 hours

There are 100 marks available for this paper.

Formulas are given on pages 164 to 166.

Section A: Pure Mathematics

1   Prove that

$$\sec 2\theta \equiv \frac{\sec^2 \theta}{2 - \sec^2 \theta}$$

(3 marks)

2    $f(x) = \dfrac{2}{x^2 - 1}$

    a)   Sketch the graph of $y = f(x)$, giving the equations of any asymptotes
        and coordinates of any points of intersection with the axes.

(4 marks)

    b)   Sketch the graph of $y = -f(x) - 1$, giving the equations of any asymptotes
        and coordinates of any points of intersection with the axes.

(3 marks)

3   The diagram shows a section of the graph of $y = 2 \cos \dfrac{x}{4}$.

    It crosses the x-axis at $x = 2\pi$ and the y-axis at $y = 2$.

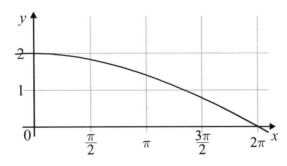

    a)   Lily works out $\displaystyle\int_0^{2\pi} 2 \cos \dfrac{x}{4} \, dx$ and gets an answer of 4π. Her teacher doesn't do the integration, but tells her that she is wrong, and that her answer is too big. Explain how the teacher knew this by looking at the graph.

(1 mark)

    b)   Find the value of $\displaystyle\int_0^{2\pi} 2 \cos \dfrac{x}{4} \, dx$.

(3 marks)

4 A group of musicians follow a 30-day programme to practise for a concert.

The length of time for which the pianist practises will increase by 10 minutes every day. On day 1 the pianist practises for 1 hour.

a) On which day will the pianist practise for exactly 4 hours 40 minutes?

(3 marks)

The length of time for which the violinist practises will increase by 4% each day. On day 1 the violinist practises for 1 hour 40 minutes.

b) On which day does the violinist first practise for more than 4 hours 40 minutes?

(4 marks)

c) Which of the two musicians practises for the longest time in total over the 30 days? Give working to show how you came to your conclusion.

(3 marks)

5 Find $\int \dfrac{45x^2 + 12x}{5x^3 + 2x^2 + 6}\, dx$.

(3 marks)

6 The curve C has parametric equations:

$$x = t^3 + 2,\ y = t^2 + 2$$

The point P lies on the curve where $t = 1$.

a) Find the coordinates of P.

(1 mark)

The line l is the tangent to C at P.

b) Find the Cartesian equation of the line l in the form $y = mx + c$.

(4 marks)

7 Find $f'(x)$ for $f(x) = \dfrac{3 + \sin x}{(2x + 1)^4}$.

(4 marks)

8   A curve has equation $x^2 - xy = 2y^3$. Find the coordinates of the point where the normal to the curve at the point $(2, 1)$ crosses the x-axis.

(7 marks)

9   **a)**   Show that the x-coordinate of the point of intersection of the curves $y = \dfrac{1}{x^3}$ and $y = x + 3$ satisfies the equation $x^4 + 3x^3 - 1 = 0$.

(1 mark)

$x^4 + 3x^3 - 1 = 0$ has a root α in the interval $[0.5, 1]$

b)   Apply the Newton-Raphson method to obtain a second and third approximation for α, using 0.6 as the first approximation for α. Give your answers to 4 decimal places.

(3 marks)

c)   Show that $\alpha = 0.650$ is correct to 3 decimal places.

(3 marks)

END OF SECTION A

Section B: Mechanics

10   A uniform rod, AB, of length 6 m and mass 6 kg rests on two supports, P and Q, as shown.

A 4 kg particle is placed on the rod at a distance of x m from A.

a)   Given that the reaction forces on the rod at P and Q are now equal, and that the rod is in equilibrium, find the value of x.

(3 marks)

b)   Another particle, of mass M kg, is added to the end of the rod, at B. Find the maximum value of M for which the rod will remain in equilibrium.

(2 marks)

11 A particle P of mass 2.5 kg is moving in a horizontal plane under the action of a single force, $\mathbf{F}$ newtons. At t seconds, the position vector of P is $\mathbf{r}$ m, where $\mathbf{r}$ is given by:

$$\mathbf{r} = (t^3 - 6t^2 + 4t)\mathbf{i} + (7t - 4t^2 + 3)\mathbf{j},$$

where $\mathbf{i}$ and $\mathbf{j}$ are the unit vectors directed due east and due north respectively.

a) Show that when $t = 5$, the velocity of P is $19\mathbf{i} - 33\mathbf{j}$.

(3 marks)

b) Find the value of t when P is moving due south.

(4 marks)

c) Find the magnitude of the resultant force acting on P when $t = 3$.

(5 marks)

12 A helicopter E is modelled as a particle. It sets off from its base O on the coast and flies at constant height with initial velocity $(40\mathbf{i} + 108\mathbf{j})$ kmh^{-1} and constant acceleration $(4\mathbf{i} + 12\mathbf{j})$ kmh^{-2}, where $\mathbf{i}$ and $\mathbf{j}$ are the unit vectors in the direction of east and north respectively. Thirty minutes after E has left the base, the pilot informs the base that he has engine failure and is ditching into the sea.

a) Find the position vector of E relative to O at the time it experiences engine failure.

(3 marks)

The base immediately informs another helicopter, F, that E is in need of assistance.
F is modelled as a particle flying towards E's position at a constant height from position $(2.5\mathbf{i} + 25.5\mathbf{j})$ km, with constant acceleration and initial velocity $(60\mathbf{i} + 120\mathbf{j})$ kmh^{-1}.

b) Given that F takes 15 minutes to reach E, find the magnitude of the acceleration of F.

(3 marks)

The base sends out a third helicopter, G, on another mission. G is modelled as a particle, and leaves the base O with velocity $(30\mathbf{i} - 40\mathbf{j})$ kmh^{-1} and accelerates at a constant $(5\mathbf{i} + 8\mathbf{j})$ kmh^{-2} towards its destination.

c) Find the position vector of G when it is travelling due east.

(6 marks)

13 A golf ball is hit off a rock of height 0.5 m with initial velocity U ms^{-1} at an angle $\alpha°$ above the horizontal. The golf ball lands on horizontal ground at point X, as shown.

a) Show that the golf ball reaches maximum height $h = \dfrac{g + U^2 \sin^2 \alpha}{2g}$ m above the ground, where g is acceleration due to gravity.

(4 marks)

b) Show that V, the speed of the golf ball when it lands at X, is $V = \sqrt{U^2 + g}$ ms^{-1}.

(5 marks)

A beach ball is projected from the same point as the golf ball, with velocity U ms^{-1} and angle of projection α.

c) Suggest two ways the model used in parts a) and b) could be adapted for the beach ball.

(2 marks)

14 Two particles, A and B, are connected by a light inextensible string that passes over a smooth pulley. A rests on a smooth plane inclined at an angle of $\theta°$ to the horizontal, where $\theta = \tan^{-1}\left(\dfrac{3}{4}\right)$, and B rests on a rough plane inclined at 70° to the horizontal, as shown. The coefficient of friction between B and the plane is 0.45. A has mass m kg and B has mass $2m$ kg.

The system is released from rest and B starts to move down the plane.

a) Find the acceleration of particle A towards the pulley.

(6 marks)

b) The mass of A is increased to 10 kg. It is now on the point of sliding down the plane. Find the mass of B.

(4 marks)

END OF EXAM PAPER

TOTAL FOR PAPER: 100 MARKS

Section One — Pure Maths

Pages 3-4: Proof

1 Proof by exhaustion: if n is even, n^3 is also even (as the product of even numbers is even), so $n^3 - n$ is even too (the difference between two even numbers is always even) *[1 mark]*.
   If n is odd, n^3 is also odd (as the product of odd numbers is odd), so $n^3 - n$ is even (as an odd number minus an odd number is even) *[1 mark]*. n is an integer so must be odd or even, so $n^3 - n$ is always even.
   You could have factorised $n^3 - n$ instead — you'd get $n(n + 1)(n - 1)$. Then if n is odd, $(n + 1)$ and $(n - 1)$ are even, so the product is even. If n is even, the product will be even too.

2 Proof by contradiction: Assume that there is a largest integer, n *[1 mark]*. Now consider a number $k = n + 1$. k is an integer, as the sum of two integers is itself an integer, and $k = n + 1 > n$ *[1 mark]*. This contradicts the assumption that n is the largest integer *[1 mark]*. Hence, there is no largest integer.

3 Take two prime numbers, p and q ($p \neq q$ and $p, q > 1$). As p is prime, its only factors are 1 and p *[1 mark]*, and as q is prime, its only factors are 1 and q *[1 mark]*. So the product pq has factors 1, p, q, and pq *[1 mark]* (these factors are found by multiplying the factors of each number together in every possible combination). $pq \neq 1$ as $p, q > 1$. Hence the product of any two distinct prime numbers has exactly four factors.

4 As a and b are rational, you can write $a = \frac{m}{n}$ and $b = \frac{p}{q}$ where m, n, p and q are integers. *[1 mark]*
   So, $c = \frac{m}{n} - \frac{p}{q} = \frac{mq - pn}{nq}$ *[1 mark]*.
   The product of two integers is always an integer and the difference of two integers is also always an integer.
   Hence, $c = \frac{mq - pn}{nq}$ is also a rational number. *[1 mark]*

5 As n is even, write n as $2k$ for some integer k *[1 mark]*.
   So $n^3 + 2n^2 + 12n = (2k)^3 + 2(2k)^2 + 12(2k)$
   $\qquad = 8k^3 + 8k^2 + 24k$ *[1 mark]*
   This can be written as $8x$, where $x = k^3 + k^2 + 3k$, so is always a multiple of 8 when n is even *[1 mark]*.

6 a) E.g. The student's proof shows that if x is even, then x^3 is even. This is not the same as showing if x^3 is even, then x is even.
      [1 mark for any sensible explanation of why the proof is not valid]

   b) Proof by contradiction:
      Assume that the statement is false. So, that means for some value of x, x^3 is even but x is odd *[1 mark]*.
      x is odd so $x = 2n + 1$, where n is an integer *[1 mark]*.
      So, $x^3 = (2n + 1)^3 = (4n^2 + 4n + 1)(2n + 1)$
      $\qquad = 8n^3 + 12n^2 + 6n + 1$
      $\qquad = 2(4n^3 + 6n^2 + 3n) + 1$ *[1 mark]*
      $4n^3 + 6n^2 + 3n$ is an integer, call it m, so $x^3 = 2m + 1$ is odd. This contradicts the assumption that x^3 is even but x is odd, *[1 mark]* hence if x^3 is even, then x is even.

7 a) E.g. Let both x and y be $\sqrt{2}$. $\sqrt{2}$ is irrational, but $\frac{\sqrt{2}}{\sqrt{2}} = 1$ which is rational, hence Riyad's claim is false.
      [1 mark for any valid counter-example]

   b) Let x be a rational number, where $x \neq 0$.
      Let z be an irrational number.
      Assume that the product zx is rational, so $zx = \frac{a}{b}$ where a and b are integers ($a, b \neq 0$) *[1 mark]*.
      x is rational, so $x = \frac{l}{m}$ where l and m are integers ($l, m \neq 0$).
      So, the product $zx = \frac{zl}{m} = \frac{a}{b} \Rightarrow z = \frac{ma}{lb}$ *[1 mark]*.
      ma and lb are the products of non-zero integers, so are non-zero integers themselves. So, z is rational. This is a contradiction, as z is irrational by definition, *[1 mark]* therefore, zx must be irrational.

Pages 5-7: Algebra and Functions 1

1 $\sqrt[m]{a^n} = a^{\frac{n}{m}}$ so $\sqrt{a^4} = a^2$ and $a^6 \times a^3 = a^{6+3} = a^9$
   So $\dfrac{a^6 \times a^3}{\sqrt{a^4}} \div a^{\frac{1}{2}} = \dfrac{a^9}{a^2} \div a^{\frac{1}{2}}$ *[1 mark]*
   $\qquad = a^{7 - \frac{1}{2}} = a^{\frac{13}{2}}$ *[1 mark]*

2 a) $3 = \sqrt[3]{27} = 27^{\frac{1}{3}}$, so $x = \frac{1}{3}$ *[1 mark]*
   b) $81 = 3^4 = (\sqrt[3]{27})^4 = 27^{\frac{4}{3}}$, so $x = \frac{4}{3}$ *[1 mark]*

3 $\dfrac{(3ab^3)^2 \times 2a^6}{6a^4 b} = \dfrac{3^2 \times a^2 \times (b^3)^2 \times 2a^6}{6a^4 b} = \dfrac{18a^8 b^6}{6a^4 b} = 3a^4 b^5$
   [2 marks available — 1 mark for simplifying the numerator to get $18a^8 b^6$, 1 mark for the correct answer]

4 $\dfrac{(5 + 4\sqrt{x})^2}{2x} = \dfrac{25 + 40\sqrt{x} + 16x}{2x}$ *[1 mark]*
   $\qquad = \frac{1}{2}x^{-1}(25 + 40x^{\frac{1}{2}} + 16x)$ *[1 mark]*
   $\qquad = \frac{25}{2}x^{-1} + 20x^{-\frac{1}{2}} + 8$ *[1 mark]*
   So $P = 20$ and $Q = 8$.

5 $(5\sqrt{5} + 2\sqrt{3})^2 = (5\sqrt{5} + 2\sqrt{3})(5\sqrt{5} + 2\sqrt{3})$
   $\qquad = (5\sqrt{5})^2 + 2(5\sqrt{5} \times 2\sqrt{3}) + (2\sqrt{3})^2$
   First term: $(5\sqrt{5})^2 = 5\sqrt{5} \times 5\sqrt{5}$
   $\qquad = 5 \times 5 \times \sqrt{5} \times \sqrt{5} = 5 \times 5 \times 5 = 125$ *[1 mark]*
   Second term: $2(5\sqrt{5} \times 2\sqrt{3}) = 2 \times 5 \times 2 \times \sqrt{5} \times \sqrt{3}$
   $\qquad = 20\sqrt{15}$ *[1 mark]*
   Third term: $(2\sqrt{3})^2 = 2\sqrt{3} \times 2\sqrt{3} = 2 \times 2 \times \sqrt{3} \times \sqrt{3}$
   $\qquad = 2 \times 2 \times 3 = 12$ *[1 mark]*
   So $(5\sqrt{5})^2 + 2(5\sqrt{5} \times 2\sqrt{3}) + (2\sqrt{3})^2$
   $\qquad = 125 + 20\sqrt{15} + 12 = 137 + 20\sqrt{15}$
   So $a = 137$, $b = 20$ and $c = 15$ *[1 mark]*.

6 Multiply top and bottom by $\sqrt{5} - 1$:
   $\dfrac{10 \times (\sqrt{5} - 1)}{(\sqrt{5} + 1) \times (\sqrt{5} - 1)}$ *[1 mark]*
   $= \dfrac{10(\sqrt{5} - 1)}{5 - 1} = \dfrac{10\sqrt{5} - 10}{4} = \dfrac{5\sqrt{5} - 5}{2} \left(= \dfrac{5(\sqrt{5} - 1)}{2} \right)$
   [1 mark for simplifying top, 1 mark for simplifying bottom]

7 Rationalise the denominator by multiplying top and bottom by $(2 - \sqrt{2})$:
   $\dfrac{4 + \sqrt{2}}{2 + \sqrt{2}} \times \dfrac{2 - \sqrt{2}}{2 - \sqrt{2}} = \dfrac{8 - 4\sqrt{2} + 2\sqrt{2} - 2}{4 - 2}$
   $\qquad = \dfrac{6 - 2\sqrt{2}}{2} = 3 - \sqrt{2}$
   [3 marks available — 1 mark for multiplying numerator and denominator by the correct expression, 1 mark for correct multiplication, 1 mark for the correct answer]

8 $\dfrac{(x^2 - 9)(3x^2 - 10x - 8)}{(6x + 4)(x^2 - 7x + 12)} = \dfrac{(x + 3)(x - 3)(3x + 2)(x - 4)}{2(3x + 2)(x - 3)(x - 4)}$
   $\qquad = \dfrac{x + 3}{2}$
   [3 marks available — 1 mark for factorising the numerator, 1 mark for factorising the denominator and 1 mark for the correct answer]

9 a) $\dfrac{x^2 + 5x - 14}{2x^2 - 4x} = \dfrac{(x + 7)(x - 2)}{2x(x - 2)} = \dfrac{x + 7}{2x}$
      [2 marks available — 1 mark for factorising the numerator and the denominator and 1 mark for cancelling to obtain correct answer]

b) $\dfrac{x^2 + 5x - 14}{2x^2 - 4x} + \dfrac{14}{x(x-4)} = \dfrac{x+7}{2x} + \dfrac{14}{x(x-4)}$

$\qquad = \dfrac{(x+7)(x-4)}{2x(x-4)} + \dfrac{2 \times 14}{2x(x-4)}$

$\qquad = \dfrac{x^2 + 3x - 28 + 28}{2x(x-4)} = \dfrac{x^2 + 3x}{2x(x-4)}$

$\qquad = \dfrac{x(x+3)}{2x(x-4)} = \dfrac{x+3}{2(x-4)}$

[3 marks available — 1 mark for putting fractions over a common denominator, 1 mark for multiplying out and simplifying the numerator and 1 mark for cancelling to obtain correct answer]

10 $\dfrac{1}{x(2x-3)} \equiv \dfrac{A}{x} + \dfrac{B}{2x-3} \Rightarrow 1 \equiv A(2x-3) + Bx$ *[1 mark]*

Equating constants gives $1 = -3A \Rightarrow A = -\dfrac{1}{3}$

Equating coefficients of x gives $0 = 2A + B \Rightarrow B = \dfrac{2}{3}$

[1 mark for A or B]

So $\dfrac{1}{x(2x-3)} \equiv \dfrac{2}{3(2x-3)} - \dfrac{1}{3x}$ *[1 mark]*

11 $\dfrac{6x-1}{x^2 + 4x + 4} = \dfrac{6x-1}{(x+2)^2}$ *[1 mark]* $\equiv \dfrac{A}{(x+2)} + \dfrac{B}{(x+2)^2}$

$\Rightarrow 6x - 1 \equiv A(x+2) + B$ *[1 mark]*

Equating coefficients of x gives $6 = A$

Equating constant terms gives $-1 = 2A + B \Rightarrow B = -13$

[1 mark for A or B]

So $\dfrac{6x-1}{x^2 + 4x + 4} \equiv \dfrac{6}{(x+2)} - \dfrac{13}{(x+2)^2}$ *[1 mark]*

Pages 8-11: Algebra and Functions 2

1 When $ax^2 + bx + c = 0$ has no real roots, you know that $b^2 - 4ac < 0$ *[1 mark]*. Here, $a = -j$, $b = 3j$ and $c = 1$.

Therefore $(3j)^2 - (4 \times -j \times 1) < 0 \Rightarrow 9j^2 + 4j < 0$ *[1 mark]*

To find the values where $9j^2 + 4j < 0$, you need to start by

solving $9j^2 + 4j = 0$: $j(9j+4) = 0$, so $j = 0$ or $9j = -4 \Rightarrow j = -\dfrac{4}{9}$

Now sketch the graph:

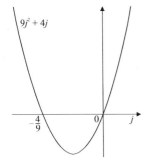

From the graph, you can see that $9j^2 + 4j < 0$ when $-\dfrac{4}{9} < j < 0$ *[1 mark]*.

2 a) Complete the square by halving the coefficient of x to find the number in the brackets (m):

$x^2 - 7x + 17 = \left(x - \dfrac{7}{2}\right)^2 + n$

$\left(x - \dfrac{7}{2}\right)^2 = x^2 - 7x + \dfrac{49}{4}$, so $n = 17 - \dfrac{49}{4} = \dfrac{19}{4}$

So $x^2 - 7x + 17 = \left(x - \dfrac{7}{2}\right)^2 + \dfrac{19}{4}$

[3 marks available — 1 mark for the correct brackets, 1 mark for finding the right correcting number and 1 mark for the correct final answer]

b) The maximum value of f(x) will be when the denominator is as small as possible — so you want the minimum value of $x^2 - 7x + 17$. Using the completed square above, you can see that the minimum value is $\dfrac{19}{4}$ *[1 mark]* because the squared part can equal but never be below 0.

So the maximum value of f(x) is $\dfrac{1}{\left(\frac{19}{4}\right)} = \dfrac{4}{19}$ *[1 mark]*.

3 If the equation has two real roots, then $b^2 - 4ac > 0$ *[1 mark]*. For this equation, $a = 3k$, $b = k$ and $c = 2$.

Use the discriminant formula to find k:

$k^2 - (4 \times 3k \times 2) > 0$ *[1 mark]*

$k^2 - 24k > 0$

$k(k - 24) > 0$

$k < 0$ or $k > 24$ *[1 mark]*

If you're struggling to solve this inequality, you could always sketch a graph like in the answer to question 1.

4 Let $y = x^3$. Then $x^6 = 7x^3 + 8$ becomes $y^2 = 7y + 8$ *[1 mark]*, so solve the quadratic in y:

$y^2 = 7y + 8 \Rightarrow y^2 - 7y - 8 = 0$

$(y - 8)(y + 1) = 0$, so $y = 8$ or $y = -1$ *[1 mark]*.

Now replace y with x^3. So $x^3 = 8 \Rightarrow x = 2$ *[1 mark]*

or $x^3 = -1 \Rightarrow x = -1$ *[1 mark]*.

Here, you had to spot that the original equation was a quadratic of the form $x^2 + bx + c$, just in terms of x^3 not x.

5 a) (i) First, rewrite the quadratic as: $-h^2 + 10h - 27$ and complete the square ($a = -1$):

$-(h - 5)^2 + 25 - 27 = -(h - 5)^2 - 2$

Rewrite the square in the form given in the question:

$T = -(-(5 - h))^2 - 2 \Rightarrow T = -(5 - h)^2 - 2$

[3 marks available — 1 mark for $(5 - h)^2$ or $(h - 5)^2$, 1 mark for $25 - 27$, 1 mark for the correct final answer]

The last couple of steps are using the fact that $(-a)^2 = a^2$ to show that $(m - n)^2 = (n - m)^2$...

(ii) $(5 - h)^2 \geq 0$ for all values of h, so $-(5 - h)^2 \leq 0$. Therefore $-(5 - h)^2 - 2 < 0$ for all h, so T is always negative *[1 mark]*.

b) (i) The maximum temperature is the maximum value of T, which is -2 (from part a) *[1 mark]*, and this occurs when the expression in the brackets $= 0$. The h-value that makes the expression in the brackets 0 is 5 *[1 mark]*, so maximum temperature occurs 5 hours after sunrise.

(ii) At sunrise, $h = 0$, so $T = 10(0) - 0^2 - 27 = -27°C$, so the graph looks like this:

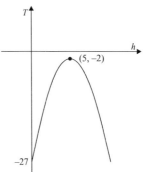

[2 marks available — 1 mark for drawing n-shaped curve that sits below the x-axis with the maximum roughly where shown (even if its position is not labelled), 1 mark for correct T-axis intercept (0, -27)]

6 Rearrange the first equation to get y on its own:

$y + x = 7 \Rightarrow y = 7 - x$ *[1 mark]*

Substitute the expression for y into the quadratic to get:

$7 - x = x^2 + 3x - 5$ *[1 mark]*

Rearrange again to get everything on one side of the equation, and then factorise it:

$0 = x^2 + 4x - 12 \Rightarrow (x + 6)(x - 2) = 0$

So $x = -6$ and $x = 2$ *[1 mark]*

Use these values to find the corresponding values of y:

When $x = -6$, $y = 7 - -6 = 13$

and when $x = 2$, $y = 7 - 2 = 5$ *[1 mark for both y-values]*

So the solutions are $x = -6$, $y = 13$ or $x = 2$, $y = 5$.

7 a) At points of intersection, $-2x + 4 = -x^2 + 3$ *[1 mark]*

$x^2 - 2x + 1 = 0$

$(x - 1)^2 = 0$ so $x = 1$ *[1 mark]*.

When $x = 1$, $y = -2x + 4 = 2$,

so there is one point of intersection at $(1, 2)$ *[1 mark]*.

b)

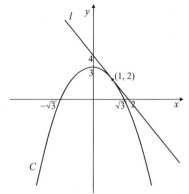

[5 marks available — 1 mark for drawing n-shaped curve,
1 mark for x-axis intercepts at ± √3 , 1 mark for maximum
point of curve and y-axis intercept at (0, 3). 1 mark for line
crossing the y-axis at (0, 4) and the x-axis at (2, 0).
1 mark for line and curve touching in one place at (1, 2).]

8 Draw the line $y = x + 2$, which has a gradient of 1, crosses the
   y-axis at $(0, 2)$ and crosses the x-axis at $(-2, 0)$. This should be a
   solid line.
   Then draw the curve $y = 4 - x^2 = (2 + x)(2 - x)$. This is an n-shaped
   quadratic which crosses the x-axis at $(-2, 0)$ and $(2, 0)$ and the
   y-axis at $(0, 4)$ (this is also the maximum point of the graph).
   This should be a dotted line.
   Then test the point $(0, 0)$ to see which side of the lines you want:
   $0 \geq 0 + 2$ — this is false, so shade the other side of the line.
   $4 - 0^2 > 0$ — this is true, so shade the region below the curve.
   So the final region (labelled R) should look like this:

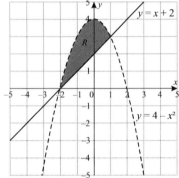

[3 marks available — 1 mark for drawing the line with correct
gradient and intercepts, 1 mark for drawing the curve with correct
intercepts, 1 mark for shading or indicating the correct region]

9 $x^2 - 8x + 15 > 0 \Rightarrow (x - 5)(x - 3) > 0$
   Sketch a graph to see where the quadratic is greater than 0 — it'll
   be a u-shaped curve that crosses the x-axis at $x = 3$ and $x = 5$.

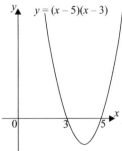

You can see from the graph that the function is positive when $x < 3$
and when $x > 5$. In set notation, this is $\{x : x < 3\} \cup \{x : x > 5\}$.
[4 marks available — 1 mark for factorising the quadratic,
1 mark for finding the roots, 1 mark for x < 3 and x > 5, 1 mark
for the correct answer in set notation]

10 a) Multiply out the brackets and rearrange to get 0 on one side:
       $(x - 1)(x^2 + x + 1) = 2x^2 - 17$
       $x^3 + x^2 + x - x^2 - x - 1 = 2x^2 - 17$ *[1 mark]*
       $x^3 - 2x^2 + 16 = 0$ *[1 mark]*

b) By the factor theorem, if $(x + 2)$ is a factor of f(x), f(−2) = 0.
    $f(x) = x^3 - 2x^2 + 16$
    $f(-2) = (-2)^3 - 2(-2)^2 + 16 = -8 - 8 + 16 = 0$ *[1 mark]*
    $f(-2) = 0$, therefore $(x + 2)$ is a factor of f(x) *[1 mark]*.

c) From part b) you know that $(x + 2)$ is a factor of f(x).
    Dividing f(x) by $(x + 2)$ gives:
    $x^3 - 2x^2 + 16 = (x + 2)(x^2 + ?x + 8) = (x + 2)(x^2 - 4x + 8)$
    If you find it easier, you can use algebraic long division here.
    [2 marks available — 2 marks for all three correct terms in
    the quadratic, otherwise 1 mark for two terms correct]

d) From b) you know that $x = -2$ is a root. From c),
    $f(x) = (x + 2)(x^2 - 4x + 8)$. So for f(x) to equal zero,
    either $(x + 2) = 0$ (so $x = -2$) or $(x^2 - 4x + 8) = 0$ *[1 mark]*.
    Completing the square of $(x^2 - 4x + 8)$ gives
    $x^2 - 4x + 8 = (x - 2)^2 + 4$, which is always positive so has no
    real roots. So $f(x) = 0$ has no solutions other than $x = -2$,
    which means it only has one root *[1 mark]*.
    You could also have shown that $x^2 - 4x + 8$ has no real roots by
    showing that the discriminant is < 0.

11 If $(2x - 1)$ is a factor of f(x), then $f\left(\frac{1}{2}\right) = 0$ by the factor theorem
    [1 mark].
    $f\left(\frac{1}{2}\right) = 2\left(\frac{1}{2}\right)^3 - 7\left(\frac{1}{2}\right)^2 - a\left(\frac{1}{2}\right) + 10$, so $0 = \frac{1}{4} - \frac{7}{4} - \frac{1}{2}a + 10$
    $\Rightarrow \frac{1}{2}a = \frac{17}{2} \Rightarrow a = 17$ *[1 mark]*.
    So $f(x) = 2x^3 - 7x^2 - 17x + 10$.
    To solve $f(x) = 0$, first factorise $2x^3 - 7x^2 - 17x + 10$. You know one
    factor, $(2x - 1)$, so find the quadratic that multiplies with that factor
    to give the original equation:
    $2x^3 - 7x^2 - 17x + 10 = (2x - 1)(x^2 + ?x - 10)$
    $\qquad\qquad\qquad\qquad = (2x - 1)(x^2 - 3x - 10)$ *[1 mark]*
    Then factorise the quadratic:
    $\qquad\qquad\qquad = (2x - 1)(x - 5)(x + 2)$ *[1 mark]*
    Again, you could've used algebraic long division here.
    Finally, solve $f(x) = 0$:
    $2x^3 - 7x^2 - 17x + 10 = 0 \Rightarrow (2x - 1)(x - 5)(x + 2) = 0$,
    so $x = \frac{1}{2}$, $x = 5$ or $x = -2$ *[2 marks for all three x-values,*
    otherwise 1 mark for either x = 5 or x = −2].

Pages 12-17: Algebra and Functions 3

1 a)

[2 marks available — 1 mark for y = |2x + 3| (with correct
x- and y-intercepts), 1 mark for y = |5x − 4| (with correct
x- and y-intercepts)]

b) From the graph, it is clear that there are two points where the
    graphs intersect. One is in the range $-\frac{3}{2} < x < \frac{4}{5}$,
    where $(2x + 3) > 0$ but $(5x - 4) < 0$.
    This gives $2x + 3 = -(5x - 4)$ *[1 mark]*.
    The other one is in the range $x > \frac{4}{5}$, where $(2x + 3) > 0$
    and $(5x - 4) > 0$, so $2x + 3 = 5x - 4$ *[1 mark]*. Solving the
    first equation gives: $2x + 3 = -5x + 4 \Rightarrow 7x = 1$, so $x = \frac{1}{7}$
    [1 mark]. Solving the second equation gives:
    $2x + 3 = 5x - 4 \Rightarrow 7 = 3x$, so $x = \frac{7}{3}$ *[1 mark]*.
    You could also have solved this algebraically by using
    $(2x + 3)^2 = (5x - 4)^2$

2 a) If $|x| = 2$, then either $x = 2$ or $x = -2$
       When $x = 2$, $|4x + 5| = |8 + 5| = |13| = 13$ *[1 mark]*
       When $x = -2$, $|4x + 5| = |-8 + 5|$ *[1 mark]* $= |-3| = 3$ *[1 mark]*

b) First, sketch a quick graph:

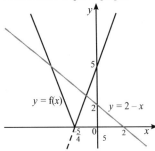

You can see that the lines cross twice, so you need to solve two inequalities:

$4x + 5 \leq 2 - x \Rightarrow 5x \leq -3 \Rightarrow x \leq -\frac{3}{5}$

and $-(4x + 5) \leq 2 - x \Rightarrow -7 \leq 3x \Rightarrow x \geq -\frac{7}{3}$

So $f(x) \leq 2 - x$ when $-\frac{7}{3} \leq x \leq -\frac{3}{5}$.

[3 marks available — 1 mark for each correct value in the inequality, 1 mark for the correct inequality signs]

Again, you could have solved this algebraically, this time by using $|x - a| \leq b \Rightarrow a - b \leq x \leq a + b$, with $a = -5$, $b = (2 - x)$ and $x = 4x$.

c) Two distinct roots means that the graphs of $y = f(x) + 2$ and $y = A$ cross twice *[1 mark]*. From the graph in part b), the graph of $y = f(x) + 2$ is the black line translated up by 2. A horizontal line will intersect this in two places as long as it lies above the point where the graph is reflected, i.e. above $y = 2$. So the possible values of A are $A > 2$ *[1 mark]*.

3 The quartic has already been factorised — there are two double roots, one at $(2, 0)$ and the other at $(-3, 0)$. When $x = 0$, $y = (-2)^2(3^2) = 36$, so the y-intercept is $(0, 36)$. The coefficient of the x^4 term is positive, and as the graph only touches the x-axis but doesn't cross it, it is always above the x-axis. The graph looks like this:

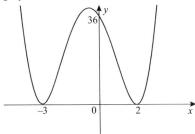

[3 marks available — 1 mark for the correct shape, 1 mark for the correct x-intercepts, 1 mark for the correct y-intercept]

4 a)

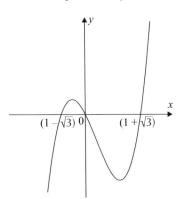

[2 marks available — 1 mark for the correct positive cubic shape with the two turning points the correct side of the y-axis and 1 mark for the x-intercepts correctly labelled.]

b) $x^3 - 2x^2 + px$ can be factorised to give $x(x^2 - 2x + p)$, so the roots of the quadratic factor must be $x = 1 + \sqrt{3}$ and $x = 1 - \sqrt{3}$. The quadratic factor can be factorised to give $(x - (1 + \sqrt{3}))(x - (1 - \sqrt{3}))$, so the constant term is given by $p = (1 + \sqrt{3})(1 - \sqrt{3}) = 1 - 3 = -2$

[2 marks available — 1 mark for a correct method to find p, 1 mark for the correct answer]

5 To transform the curve $y = x^3$ into $y = (x - 1)^3$, translate it 1 unit horizontally to the right (in the positive x-direction) *[1 mark]*. To transform this into the curve $y = 2(x - 1)^3$, stretch it vertically (parallel to the y-axis) by a scale factor of 2 *[1 mark]*. Finally, to transform into the curve $y = 2(x - 1)^3 + 4$, the whole curve is translated 4 units upwards (in the positive y-direction) *[1 mark]*.

6 a) Expand the brackets to show the two functions are the same:
$$(2t + 1)(t - 2)(t - 3.5) = (2t + 1)(t^2 - 5.5t + 7)$$
$$= 2t^3 - 11t^2 + 14t + t^2 - 5.5t + 7$$
$$= 2t^3 - 10t^2 + 8.5t + 7 \text{ as required}$$

[2 marks available — 1 mark for expanding the brackets, 1 mark for rearranging to get the required answer]

b)

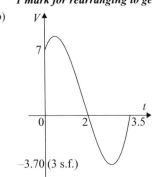

[3 marks available — 1 mark for the correct shape drawn between t = 0 and t = 3.5, 1 mark for the correct t-intercepts, 1 mark for the correct V-intercept]

c) 2 s *[1 mark]*
This is the first point on the graph at which $V = 0$.

d) When the diver starts his dive (i.e. $t = 0$), $V = 7$. The diver's height is 1.75 m, so the diving board is $7 - 1.75 = 5.25$ m high.
[2 marks available — 1 mark for a correct method, 1 mark for the correct answer]

e) The adapted model is a vertical translation of the original graph by 3 m upwards. This means that the lowest point of the dive is $3.70 - 3 = 0.70$ m below the surface of the pool. This is obviously unrealistic as it is far too shallow — you would expect the diver to go at least as deep as from the lower diving board, so the adapted model is not valid.
[2 marks available — 1 mark for stating that the model is not valid, 1 mark for a sensible explanation of why]
This model is a lot easier to comment on if you realise it's just a vertical translation of the original model.

7 a)

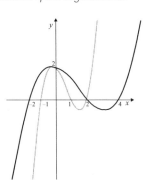

[3 marks available — 1 mark for horizontal stretch, 1 mark for x-axis intercepts at –2, 2 and 4, 1 mark for correct y-axis intercept at 2]

b)

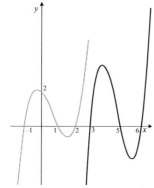

[3 marks available — 1 mark for vertical stretch, 1 mark for horizontal translation to the right, 1 mark for x-axis intercepts at 3, 5 and 6]

8 a)

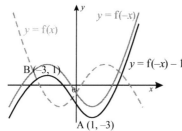

[3 marks available — 1 mark for shape (reflection in the y-axis and translation), 1 mark each for coordinates of A and B after transformation]

The solid grey line shows the graph of y = f(−x) — it's easier to do the transformation in two stages, instead of doing it all at once.

b)

[3 marks available — 1 mark for shape (stretch and translation), 1 mark each for coordinates of A and B after transformation]

9 a) A translation of 3 up is a translation of the form f(x) + 3 *[1 mark]*, then a translation of 2 right is f(x − 2) + 3 *[1 mark]*. Finally, a reflection in the y-axis gives f(−(x − 2)) + 3. So g(x) = f(2 − x) + 3 *[1 mark]*.

b) Original coordinates of P: (1, 2).
After translation of 3 up and 2 right: (3, 5)
After reflection in the y-axis: (−3, 5) *[1 mark]*.
Original coordinates of Q: (3, 16).
After translation of 3 up and 2 right: (5, 19)
After reflection in the y-axis: (−5, 19) *[1 mark]*.
Do a quick sketch of the graph if you need to.

10 a) (i) gf(x) = g(2^x) = $\sqrt{3(2^x) + 1}$ *[1 mark]*

(ii) gf(x) = 5 ⇒ $\sqrt{3(2^x) + 1}$ = 5 ⇒ $3(2^x) + 1 = 25$
⇒ $3(2^x) = 24$ ⇒ $2^x = 8$ ⇒ x = 3
[2 marks available — 1 mark for a correct method, 1 mark for the correct answer]

b) First write y = g(x) and rearrange to make x the subject:
y = $\sqrt{3x + 1}$ → $y^2 = 3x + 1$ ⇒ $y^2 − 1 = 3x$
⇒ $\frac{y^2 − 1}{3} = x$
Then replace x with $g^{-1}(x)$ and y with x:
$g^{-1}(x) = \frac{x^2 − 1}{3}$ *[1 mark]*
g(x) has domain $x \geq -\frac{1}{3}$ and range g(x) ≥ 0, so $g^{-1}(x)$ has domain x ≥ 0 *[1 mark]* and range $g^{-1}(x) \geq -\frac{1}{3}$ *[1 mark]*.

11 a) g has range g(x) ≥ −k *[1 mark]*, as the minimum value of g is −k.

b) Neither f nor g are one-to-one functions, so they don't have inverses *[1 mark]*.

c) (i) gf(1) = g$\left(\frac{1}{1^2}\right)$ = g(1) = $1^2 − k$ *[1 mark]*
So 1 − k = −8 ⇒ k = 9 *[1 mark]*
fg(x) = f($x^2 − 9$) *[1 mark]* = $\frac{1}{(x^2 − 9)^2}$ *[1 mark]*.
The domain of fg is $x \in \mathbb{R}, x \neq \pm 3$ *[1 mark]*, as the denominator of the function can't be 0.

(ii) From part (i), you know that fg(x) = $\frac{1}{(x^2 − 9)^2}$, so
$\frac{1}{(x^2 − 9)^2} = \frac{1}{256}$ ⇒ $(x^2 − 9)^2 = 256$ *[1 mark]*
$x^2 − 9 = \pm\sqrt{256} = \pm 16$ *[1 mark]*
$x^2 = 9 \pm 16 = 25, −7$ *[1 mark]*
$x = \sqrt{25} = \pm 5$ *[1 mark]*
You can ignore $x^2 = −7$, as this has no solutions in $x \in \mathbb{R}$.

Pages 18-22: Coordinate Geometry

1 a) To find the coordinates of A, solve the equations of the lines simultaneously:
l_1: x − y + 1 = 0
l_2: 2x + y − 8 = 0
Add the equations to get rid of y:
3x − 7 = 0 *[1 mark]* ⇒ x = $\frac{7}{3}$ *[1 mark]*
Now put x = $\frac{7}{3}$ back into l_1 to find y:
$\frac{7}{3}$ − y + 1 = 0 ⇒ y = $\frac{7}{3}$ + 1 = $\frac{10}{3}$
So A is $\left(\frac{7}{3}, \frac{10}{3}\right)$ *[1 mark]*

b) There's a lot of information here, so draw a quick sketch to make things a bit clearer:

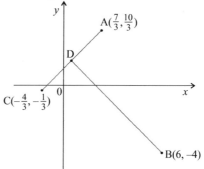

To find the equation of the line through B and D, you need its gradient. But before you can find the gradient, you need to find the coordinates of point D — the midpoint of AC. To find the midpoint of two points, find the average of the x-values and the average of the y-values:

D = $\left(\frac{x_A + x_C}{2}, \frac{y_A + y_C}{2}\right) = \left(\frac{\frac{7}{3} + \frac{−4}{3}}{2}, \frac{\frac{10}{3} + \frac{−1}{3}}{2}\right)$ *[1 mark]*

D = $\left(\frac{1}{2}, \frac{3}{2}\right)$ *[1 mark]*

To find the gradient (m) of the line through B and D, use this rule: $m_{BD} = \frac{y_D − y_B}{x_D − x_B}$

$m = \frac{\frac{3}{2} − −4}{\frac{1}{2} − 6} = \frac{\frac{3}{2} + \frac{8}{2}}{\frac{1}{2} − \frac{12}{2}} = \frac{3 + 8}{1 − 12} = −1$ *[1 mark]*

Now you can find the equation of the line. Input the known values of x and y at B(6, −4) and the gradient (−1) into y − y_1 = m(x − x_1), which gives:
y − (−4) = −1(x − 6) *[1 mark]*
⇒ y + 4 = −x + 6 ⇒ x + y − 2 = 0 *[1 mark]*
You could also have used the point D you found earlier in the question in the formula y − y_1 = m(x − x_1).

c) Look at the sketch in part b). To prove triangle ABD is a right-angled triangle, you need to prove that lines AD and BD are perpendicular — in other words, prove the product of their gradients equals −1.

You already know the gradient of BD = −1.
Use the same rule to find the gradient of AD:

$$m_{AD} = \frac{y_D - y_A}{x_D - x_A}$$

$$m = \frac{\frac{3}{2} - \frac{10}{3}}{\frac{1}{2} - \frac{7}{3}} = \frac{\frac{9}{6} - \frac{20}{6}}{\frac{3}{6} - \frac{14}{6}} = \frac{9-20}{3-14} = 1 \text{ [1 mark]}$$

$m_{BD} \times m_{AD} = -1 \times 1 = -1$ **[1 mark]**, so triangle ABD is a right-angled triangle **[1 mark]**.

2 a) The gradient of the line through B and C equals the coefficient of x when the equation of the line is in the form $y = mx + c$.
$-3x + 5y = 16 \Rightarrow 5y = 3x + 16 \Rightarrow y = \frac{3}{5}x + \frac{16}{5}$,
so gradient $= \frac{3}{5}$.
AB and BC are perpendicular, so the gradient of the line through A and B equals $-1 \div \frac{3}{5} = -\frac{5}{3}$ **[1 mark]**.
Two lines are parallel if they have the same gradient.
$5x + 3y - 6 = 0 \Rightarrow 3y = -5x + 6 \Rightarrow y = -\frac{5}{3}x + 2$,
so the gradient is $-\frac{5}{3}$ **[1 mark]**.
The line with equation $5x + 3y - 6 = 0$ has the same gradient as the line through points A and B, so the lines are parallel **[1 mark]**.

b) To calculate the area, find the length of one side — say AB. Point B has coordinates $(3, k)$, so you can find k by substituting $x = 3$ and $y = k$ into the equation of the line through B and C:
$-3x + 5y = 16 \Rightarrow (-3 \times 3) + 5k = 16 \Rightarrow 5k = 25 \Rightarrow k = 5$,
so point B = $(3, 5)$ **[1 mark]**. Now you can input the values of x and y at B(3, 5) and the gradient $(-\frac{5}{3})$ into $y - y_1 = m(x - x_1)$ to find the equation of AB:
$y - 5 = -\frac{5}{3}(x - 3) \Rightarrow y = -\frac{5}{3}x + 10$ **[1 mark]**
Use this to find the coordinates of point A:
A lies on the y-axis, so $x = 0$.
When $x = 0$, $y = 10$, so A is the point (0, 10) **[1 mark]**.
Now find the length AB using Pythagoras' theorem:
$(AB)^2 = (10 - 5)^2 + (0 - 3)^2$ **[1 mark]** $= 25 + 9 = 34$,
Area of square = $(AB)^2 = 34$ units^2 **[1 mark]**

3 a) The centre of the circle must be the midpoint of AB, since AB is a diameter. Midpoint of AB is:
$\left(\frac{2+0}{2}, \frac{1+-5}{2}\right) = (1, -2)$ **[1 mark]**
The radius is the distance from the centre (1, −2) to point A:
radius $= \sqrt{(2-1)^2 + (1-(-2))^2}$ **[1 mark]** $= \sqrt{10}$ **[1 mark]**

b) The general equation for a circle with centre (a, b) and radius r is: $(x - a)^2 + (y - b)^2 = r^2$. So for a circle with centre (1, −2) and radius $\sqrt{10}$, that gives $(x - 1)^2 + (y + 2)^2 = 10$ **[1 mark]**.
To show that the point (4, −1) lies on the circle, show that it satisfies the equation of the circle: $(4 - 1)^2 + (-1 + 2)^2$ $= 9 + 1 = 10$, so (4, −1) lies on the circle **[1 mark]**.

c) Start with your equation from part b) and multiply out to get the form given in the question:
$(x - 1)^2 + (y + 2)^2 = 10$
$x^2 - 2x + 1 + y^2 + 4y + 4 = 10$
$x^2 + y^2 - 2x + 4y - 5 = 0$
[2 marks available — 1 mark for multiplying out the equation from part b, 1 mark for correct rearrangement to give the answer in the required form]

d) The radius at A has the same gradient as the diameter AB, so gradient of radius $= \frac{1 - -5}{2 - 0} = 3$ **[1 mark]**.
The tangent at A is perpendicular to the radius at point A, so the tangent has gradient $-1 \div 3 = -\frac{1}{3}$ **[1 mark]**.

Put the gradient $-\frac{1}{3}$ and point A(2, 1) into the formula for the equation of a straight line and rearrange:
$y - y_1 = m(x - x_1) \Rightarrow y - 1 = -\frac{1}{3}(x - 2)$
$\Rightarrow y - 1 = -\frac{1}{3}x + \frac{2}{3} \Rightarrow y = -\frac{1}{3}x + \frac{5}{3}$ **[1 mark]**

4 a) The line through the centre P bisects the chord, and so is perpendicular to the chord AB at the midpoint M.
Gradient of AB = Gradient of AM = $\frac{(7 - 10)}{(11 - 9)} = -\frac{3}{2}$.
Gradient of PM = $-1 \div -\frac{3}{2} = \frac{2}{3}$
Gradient of PM = $\frac{(7 - 3)}{(11 - p)} = \frac{2}{3}$
$\Rightarrow 3(7 - 3) = 2(11 - p) \Rightarrow 12 = 22 - 2p \Rightarrow p = 5$.
[5 marks available — 1 mark for identifying that PM and AB are perpendicular, 1 mark for correct gradient of AB (or AM), 1 mark for correct gradient of PM, 1 mark for substitution of the y-coordinate of P into the equation for the gradient or equation of the line PM, and 1 mark for correct rearrangement to give the answer in the required form]

b) The equation of a circle is $(x - a)^2 + (y - b)^2 = r^2$.
The centre of the circle is P(5, 3), so $a = 5$ and $b = 3$. **[1 mark]**
r^2 is the square of the radius. The radius equals the length of AP, so you can find r^2 using Pythagoras' theorem:
$r^2 = (AP)^2 = (9 - 5)^2 + (10 - 3)^2$ **[1 mark]** $= 65$
So the equation of the circle is:
$(x - 5)^2 + (y - 3)^2 = 65$ **[1 mark]**

5 a) A is on the y-axis, so the x-coordinate is 0.
Just put $x = 0$ into the equation and solve:
$0^2 - (6 \times 0) + y^2 - 4y = 0$ **[1 mark]**
$\Rightarrow y^2 - 4y = 0 \Rightarrow y(y - 4) = 0 \Rightarrow y = 0$ or $y = 4$
$y = 0$ is the origin, so A is at (0, 4) **[1 mark]**

b) Complete the square for the terms involving x and y separately:
Completing the square for $x^2 - 6x$ means you have to start with $(x - 3)^2$, but $(x - 3)^2 = x^2 - 6x + 9$, so you need to subtract 9:
$(x - 3)^2 - 9$ **[1 mark]**
Now the same for $y^2 - 4y$: $(y - 2)^2 = y^2 - 4y + 4$,
so subtract 4 which gives: $(y - 2)^2 - 4$ **[1 mark]**
Put these new expressions back into the original equation:
$(x - 3)^2 - 9 + (y - 2)^2 - 4 = 0$
$\Rightarrow (x - 3)^2 + (y - 2)^2 = 13$ **[1 mark]**

c) In the general equation for a circle $(x - a)^2 + (y - b)^2 = r^2$, the centre is (a, b) and the radius is r.
So for the equation in part b), $a = 3$, $b = 2$, $r = \sqrt{13}$
Hence, the centre is (3, 2) **[1 mark]**
and the radius is $\sqrt{13}$ **[1 mark]**

d) The tangent at point A is perpendicular to the radius at A.
The radius between A(0, 4) and the centre(3, 2) has gradient:
$\frac{y_2 - y_1}{x_2 - x_1} = \frac{2 - 4}{3 - 0} = -\frac{2}{3}$ **[1 mark]**
So the gradient of the tangent at A $= -1 \div -\frac{2}{3} = \frac{3}{2}$ **[1 mark]**
Put $m = \frac{3}{2}$ and A = (0, 4) into $y - y_1 = m(x - x_1)$ to find the equation of the tangent to the circle at point A:
$y - 4 = \frac{3}{2}(x - 0) \Rightarrow y - 4 = \frac{3}{2}x \Rightarrow y = \frac{3}{2}x + 4$ **[1 mark]**
Point B lies on the line with equation $y = \frac{3}{2}x + 4$.
B also lies on the x-axis, so substitute $y = 0$ into the equation of the line to find the x-coordinate of B:
$0 = \frac{3}{2}x + 4 \Rightarrow x = -\frac{8}{3}$, so B is the point $\left(-\frac{8}{3}, 0\right)$ **[1 mark]**
Now find AB using Pythagoras' theorem:
$(AB)^2 = \left(0 - -\frac{8}{3}\right)^2 + (4 - 0)^2$ **[1 mark]** $= \frac{64}{9} + 16$,
so $AB = \sqrt{\frac{64}{9} + 16} = \sqrt{\frac{208}{9}} = \frac{4\sqrt{13}}{3}$ **[1 mark]**

If the question asks for an exact answer, leave it in surd form.

6   First, find the centre of the circle. Do this by finding the perpendicular bisectors of any two sides:

Midpoint of $PQ = \left(\frac{3+0}{2}, \frac{1+2}{2}\right) = \left(\frac{3}{2}, \frac{3}{2}\right)$,

gradient of $PQ = \frac{y_P - y_Q}{x_P - x_Q} = \frac{1-2}{3-0} = \frac{-1}{3} = -\frac{1}{3}$

So the perpendicular bisector of PQ has

gradient $-1 \div -\frac{1}{3} = 3$ *[1 mark]* and goes through $\left(\frac{3}{2}, \frac{3}{2}\right)$, so has

equation $y - \frac{3}{2} = 3\left(x - \frac{3}{2}\right) \Rightarrow y = 3x - 3$ *[1 mark]*.

Midpoint of $PR = \left(\frac{3+1}{2}, \frac{1+5}{2}\right) = (2, 3)$,

gradient of $PR = \frac{y_P - y_R}{x_P - x_R} = \frac{1-5}{3-1} = \frac{-4}{2} = -2$

So the perpendicular bisector of PR has gradient $-1 \div -2 = \frac{1}{2}$
[1 mark] and goes through $(2, 3)$, so has equation

$y - 3 = \frac{1}{2}(x - 2) \Rightarrow y = \frac{1}{2}x + 2$ *[1 mark]*.

Find the centre of the circle by setting these equations of the perpendicular bisectors equal to one another and solving:

$3x - 3 = \frac{1}{2}x + 2 \Rightarrow \frac{5}{2}x = 5 \Rightarrow x = 2$, so $y = (3 \times 2) - 3 = 3$

So the centre is $(2, 3)$ *[1 mark]*.

The distance from the centre to point Q $(0, 2)$ is the radius:
$r = \sqrt{(2-0)^2 + (3-2)^2} = \sqrt{4+1} = \sqrt{5}$, so $r^2 = (\sqrt{5})^2 = 5$ *[1 mark]*
So the equation of C is $(x - 2)^2 + (y - 3)^2 = 5$ *[1 mark]*.
There are lots of different ways to get the right answer here, depending on which sides or points you use in your calculation.

7   a)   At A, $y = 4$, so $4 \cos \theta = 4 \Rightarrow \cos \theta = 1$

$\Rightarrow \theta = 0$, as $0 \le \theta \le \frac{\pi}{2}$.

At B, $x = 3$, so $3 \sin \theta = 3 \Rightarrow \sin \theta = 1$

$\Rightarrow \theta = \frac{\pi}{2}$, as $0 \le \theta \le \frac{\pi}{2}$.
[2 marks available — 1 mark for each value of θ]

b) $y^2 = 16 \cos^2 \theta$. Use the identity $\cos^2 \theta \equiv 1 - \sin^2 \theta$:
$y^2 = 16(1 - \sin^2 \theta)$ *[1 mark]*.

Now, $x^2 = 9 \sin^2 \theta$, so $\frac{x^2}{9} = \sin^2 \theta$. *[1 mark]*
Substitute this into the equation for y^2:

$y^2 = 16\left(1 - \frac{x^2}{9}\right) = 16 - \frac{16x^2}{9} = \left(4 + \frac{4x}{3}\right)\left(4 - \frac{4x}{3}\right)$ *[1 mark]*
The last step is a difference of squares.

8   For the boat to be further west than the tip of the island, this means $x < 12$, so:
$t^2 - 7t + 12 < 12$ *[1 mark]* $\Rightarrow t^2 - 7t < 0 \Rightarrow t(t - 7) < 0$ *[1 mark]*
This means that the boat starts level with the tip of the island (at $t = 0$), and is then level again when $t = 7$. So the boat is west of the tip of the island for 7 hours *[1 mark]*.

9   a)   Substitute the given value of θ into the parametric equations:

$\theta = \frac{\pi}{3} \Rightarrow x = 1 - \tan\frac{\pi}{3} = 1 - \sqrt{3}$

$y = \frac{1}{2}\sin\left(\frac{2\pi}{3}\right) = \frac{1}{2}\left(\frac{\sqrt{3}}{2}\right) = \frac{\sqrt{3}}{4}$

So $P = \left(1 - \sqrt{3}, \frac{\sqrt{3}}{4}\right)$

[2 marks available — 1 mark for substituting $\theta = \frac{\pi}{3}$ into the parametric equations, 1 mark for both coordinates of P correct]

b)   Use $y = -\frac{1}{2}$ to find the value of θ:

$-\frac{1}{2} = \frac{1}{2}\sin 2\theta \Rightarrow \sin 2\theta = -1$

$\Rightarrow 2\theta = -\frac{\pi}{2} \Rightarrow \theta = -\frac{\pi}{4}$

[2 marks available — 1 mark for substituting given x- or y-value into the correct parametric equation, 1 mark for finding the correct value of θ]
You can also find θ using the parametric equation for x, with $x = 2$.

c) $x = 1 - \tan\theta \Rightarrow \tan\theta = 1 - x$

$y = \frac{1}{2}\sin 2\theta = \frac{1}{2}\left(\frac{2\tan\theta}{1 + \tan^2\theta}\right)$

$= \frac{\tan\theta}{1 + \tan^2\theta} = \frac{(1-x)}{1 + (1-x)^2} = \frac{1 - x}{1 + 1 - 2x + x^2}$

$= \frac{1 - x}{x^2 - 2x + 2}$

[3 marks available — 1 mark for using the given identity to rearrange one of the parametric equations, 1 mark for eliminating θ from the parametric equation for y, 1 mark for correctly expanding to obtain the Cartesian equation given in the question]

10 a) C crosses the y-axis when $x = 0$,
so when $4t - 2 = 0 \Rightarrow 4t = 2 \Rightarrow t = \frac{1}{2}$ *[1 mark]*
Substitute this into the equation for y:

$y = \left(\frac{1}{2}\right)^3 + \frac{1}{2} = \frac{5}{8}$, so the coordinates are $\left(0, \frac{5}{8}\right)$ *[1 mark]*

b)   Substitute $x = 4t - 2$ into $y = \frac{1}{2}x + 1$:

$y = \frac{1}{2}(4t - 2) + 1 = 2t$. Now solve for t when $y = t^3 + t$:
$2t = t^3 + t \Rightarrow t^3 - t = 0 \Rightarrow t(t^2 - 1) = 0 \Rightarrow t(t-1)(t+1) = 0$.
So $t = 0$, 1 and -1.
When $t = 0$, $x = -2$ and $y = 0$ so the point has coordinates $(-2, 0)$
When $t = 1$, $x = 2$ and $y = 2$ so the point has coordinates $(2, 2)$
When $t = -1$, $x = -6$ and $y = -2$ so the point
has coordinates $(-6, -2)$
[4 marks available — 1 mark for a correct method to find t at points of intersection, 1 mark for all values of t correct, 2 marks for all coordinates correct, otherwise 1 mark for two coordinates correct]

c) $x = 4t - 2 \Rightarrow t = \frac{x + 2}{4}$ *[1 mark]*
Substitute this into the equation for y:

$y = \left(\frac{x+2}{4}\right)^3 + \frac{x+2}{4} = \frac{x^3 + 6x^2 + 12x + 8}{64} + \frac{x+2}{4}$ *[1 mark]*

$= \frac{x^3 + 6x^2 + 28x + 40}{64} = \frac{1}{64}x^3 + \frac{3}{32}x^2 + \frac{7}{16}x + \frac{5}{8}$ *[1 mark]*

(so $a = \frac{1}{64}$, $b = \frac{3}{32}$, $c = \frac{7}{16}$ and $d = \frac{5}{8}$)

Pages 23-25: Sequences and Series 1

1   a)   In an arithmetic series, the nth term is defined by the formula $a + (n - 1)d$. The 12th term is 79, so the equation is
$79 = a + 11d$, and the 16th term is 103, so the other equation
is $103 = a + 15d$ *[1 mark for both equations]*. Solving these simultaneously (by taking the first equation away from the second) gives $24 = 4d$, so $d = 6$ *[1 mark]*.
Putting this value of d into the first equation gives
$79 = a + (11 \times 6)$, so $a = 13$. *[1 mark]*

b) $S_n = \frac{n}{2}[2a + (n - 1)d]$
Putting in the values of a and d from above, and $n = 15$:

$S_{15} = \frac{15}{2}[26 + 14 \times 6]$ *[1 mark]* $= 825$ *[1 mark]*
The formula for S_n is given on the formula pages.

2   a)   Work out the terms one by one:
$x_1 = k$
$x_2 = 3k - 4$
$x_3 = 3(3k - 4) - 4 = 9k - 16$ *[1 mark]*
$x_4 = 3(9k - 16) - 4 = 27k - 52$ *[1 mark]*

b)   Using the terms above with $k = 1$, the sequence is:
$1, -1, -7, -25, ...$ *[1 mark]*
As the sequence involves multiplying the previous number by three and then taking away a number, once it is negative, it will only decrease (it'll become a larger negative number), therefore the sequence is decreasing *[1 mark]*.

3   a) $h_2 = 2h_1 + 2 = 2 \times 5 + 2 = 12$
$h_3 = 2h_2 + 2 = 2 \times 12 + 2 = 26$
$h_4 = 2h_3 + 2 = 2 \times 26 + 2 = 54$
[2 marks available — 1 mark for a correct method, 1 mark for all three correct answers]

b) $\sum_{r=3}^{6} h_r = h_3 + h_4 + h_5 + h_6$

$h_5 = 2h_4 + 2 = 2 \times 54 + 2 = 110$ and

$h_6 = 2h_5 + 2 = 2 \times 110 + 2 = 222$ *[1 mark for both correct]*

So $\sum_{r=3}^{6} h_r = 26 + 54 + 110 + 222$ *[1 mark]* $= 412$ *[1 mark]*

4 a) First put the two known terms into the formula for the nth term of a geometric series, $u_n = ar^{n-1}$:

$u_3 = ar^2 = \frac{5}{2}$ and $u_6 = ar^5 = \frac{5}{16}$ *[1 mark for both]*

Divide the expression for u_6 by the expression for u_3 to get an expression just containing r and solve it:

$\frac{ar^5}{ar^2} = \frac{5}{16} \div \frac{5}{2} \Rightarrow r^3 = \frac{5}{16} \times \frac{2}{5} = \frac{10}{80}$

$r^3 = \frac{1}{8} \Rightarrow r = \sqrt[3]{\frac{1}{8}} \Rightarrow r = \frac{1}{2}$ *[1 mark]*

Put this value back into the expression for u_3 to find a:

$a\left(\frac{1}{2}\right)^2 = \frac{5}{2} \Rightarrow \frac{a}{4} = \frac{5}{2} \Rightarrow a = 10$ *[1 mark]*

The nth term is $u_n = ar^{n-1}$, where $r = \frac{1}{2}$ and $a = 10$

$u_n = 10 \times \left(\frac{1}{2}\right)^{n-1} = 10 \times \frac{1}{2^{n-1}} = \frac{10}{2^{n-1}}$ *[1 mark]*

b) $S_n = \frac{a(1 - r^n)}{1 - r}$,

$S_{10} = \frac{10\left(1 - \left(\frac{1}{2}\right)^{10}\right)}{1 - \left(\frac{1}{2}\right)}$ *[1 mark]* $= 10 \times 2 \times \left(1 - \frac{1}{2^{10}}\right)$

$= 20\left(1 - \frac{1}{1024}\right) = \frac{5115}{256}$ *[1 mark]*

c) Substitute $a = 10$ and $r = \frac{1}{2}$ into the sum to infinity formula:

$S_\infty = \frac{a}{1 - r} = \frac{10}{1 - \left(\frac{1}{2}\right)} = 10 \div \frac{1}{2} = 10 \times 2 = 20$

[2 marks available — 1 mark for a correct method and 1 mark for showing the sum to infinity is 20]

5 a) The series is defined by $u_{n+1} = 5 \times 1.7^n$ so $r = 1.7$.
r is greater than 1, so the sequence is divergent, which means the sum to infinity cannot be found *[1 mark]*.

b) $u_3 = 5 \times 1.7^2 = 14.45$ *[1 mark]*
$u_8 = 5 \times 1.7^7 = 205.17$ (2 d.p.) *[1 mark]*

6 a) $S_\infty = \frac{a}{1 - r} = \frac{20}{1 - \frac{3}{4}} = \frac{20}{\frac{1}{4}} = 80$

[2 marks available — 1 mark for substituting into the correct formula, 1 mark for correct answer]

b) $u_{15} = ar^{14} = 20 \times \left(\frac{3}{4}\right)^{14} = 0.356$ (to 3 s.f.)

[2 marks available — 1 mark for substituting into the correct formula, 1 mark for correct answer]

c) Use the formula for the sum of a geometric series to write an expression for S_n:

$S_n = \frac{a(1 - r^n)}{1 - r} = \frac{20\left(1 - \left(\frac{3}{4}\right)^n\right)}{1 - \frac{3}{4}}$ *[1 mark]*

so $\frac{20\left(1 - \left(\frac{3}{4}\right)^n\right)}{1 - \frac{3}{4}} > 79.76$

Now rearrange and use logs to get n on its own:

$\frac{20\left(1 - \left(\frac{3}{4}\right)^n\right)}{1 - \frac{3}{4}} > 79.76 \Rightarrow 20\left(1 - \left(\frac{3}{4}\right)^n\right) > 19.94$

$\Rightarrow 1 - \left(\frac{3}{4}\right)^n > 0.997 \Rightarrow 0.003 > 0.75^n$ *[1 mark]*

$\Rightarrow \log 0.003 > n \log 0.75$ *[1 mark]*

$\Rightarrow \frac{\log 0.003}{\log 0.75} < n$ *[1 mark]*

Remember — if x < 1, then log x has a negative value and dividing by a negative means flipping the inequality.

$\frac{\log 0.003}{\log 0.75} = 20.1929...$

so $n > 20.1929...$

But n must be an integer, so $n = 21$ *[1 mark]*

7 a) Use $u_n = ar^{n-1}$ with $a = 1$ and $r = 1.5$:
$u_5 = 1 \times (1.5)^4$ *[1 mark]*
$= 5.06$ km (to the nearest 10 m) *[1 mark]*

b) Use $u_n = ar^{n-1}$ with $a = 2$ and $r = 1.2$:
$u_9 = 2 \times (1.2)^8 = 8.60$ km (3 s.f.) *[1 mark]*
$u_{10} = 2 \times (1.2)^9 = 10.3$ km (3 s.f.) *[1 mark]*
$u_9 < 10$ km and $u_{10} > 10$ km, so day 10 is the first day that Chris runs further than 10 km. *[1 mark]*
You could've used logs to solve $2(1.2)^{n-1} > 10$ here instead.

c) Alex: $3 \times 10 = 30$ km *[1 mark]*

Use the formula for the sum of first n terms: $S_n = \frac{a(1 - r^n)}{1 - r}$

Chris: $\frac{2(1 - 1.2^{10})}{1 - 1.2} = 51.917...$ km *[1 mark]*

Heather: $\frac{1(1 - 1.5^{10})}{1 - 1.5} = 113.330...$ km *[1 mark]*

Total raised $= 30 + 51.917... + 113.330...$
$= £195.25$ (to the nearest penny) *[1 mark]*

8 a) $S_\infty = \frac{a}{1 - r} = -9 \Rightarrow a = -9(1 - r)$

and $ar = -2 \Rightarrow a = \frac{-2}{r}$

$\Rightarrow \frac{-2}{r} = -9(1 - r)$

$\Rightarrow -2 = -9r + 9r^2 \Rightarrow 9r^2 - 9r + 2 = 0$

[3 marks available — 1 mark for finding two expressions in a and r, 1 mark for setting these expressions equal to each other, 1 mark for rearranging to give answer in required form]

b) $9r^2 - 9r + 2 = 0 \Rightarrow (3r - 1)(3r - 2) = 0$ *[1 mark]*

$\Rightarrow r = \frac{1}{3}$ or $r = \frac{2}{3}$ *[1 mark for both]*

c) $ar = -2 \Rightarrow a = \frac{-2}{r}$

$r = \frac{1}{3} \Rightarrow a = \frac{-2}{\frac{1}{3}} = -6$ *[1 mark]*

$r = \frac{2}{3} \Rightarrow a = \frac{-2}{\frac{2}{3}} = -3$ *[1 mark]*

Pages 26-29: Sequences and Series 2

1 a) Expand $(1 + ax)^{10}$ using the binomial expansion formula:

$(1 + ax)^{10} = 1 + \frac{10}{1}(ax) + \frac{10 \times 9}{1 \times 2}(ax)^2 + \frac{10 \times 9 \times 8}{1 \times 2 \times 3}(ax)^3 + ...$

$= 1 + 10ax + 45a^2x^2 + 120a^3x^3$

[2 marks available — 1 mark for substituting into the formula correctly, 1 mark for correct answer]

b) First take out a factor of 2 to get it in the form $(1 + ax)^n$:

$(2 + 3x)^5 = \left[2\left(1 + \frac{3}{2}x\right)\right]^5 = 2^5\left(1 + \frac{3}{2}x\right)^5 = 32\left(1 + \frac{3}{2}x\right)^5$

[1 mark]

Now expand: $32\left[1 + \frac{5}{1}\left(\frac{3}{2}x\right) + \frac{5 \times 4}{1 \times 2}\left(\frac{3}{2}x\right)^2 + ...\right]$

You only need the x^2 term, so simplify that one:

$32 \times \frac{5 \times 4}{1 \times 2} \times \left(\frac{3}{2}\right)^2 \times x^2$ *[1 mark]* $= 720x^2$

So the coefficient of x^2 is 720 *[1 mark]*

c) Look back to the x^2 term from part a) — it's $45a^2x^2$.
This is equal to $720x^2$ so just rearrange the equation to find a:
$45a^2 = 720 \Rightarrow a^2 = 16 \Rightarrow a = \pm 4$
From part a), $a > 0$, so $a = 4$ *[1 mark]*

2 a) c represents the coefficient of x^3, so find an expression for the coefficient of x^3 using the binomial expansion formula:

$(j + kx)^6 = j^6\left(1 + \frac{k}{j}x\right)^6$

Coefficient of $x^3 = j^6 \times \frac{6 \times 5 \times 4}{1 \times 2 \times 3} \times \left(\frac{k}{j}\right)^3$ *[1 mark]*

so $j^6 \times \frac{6 \times 5 \times 4}{1 \times 2 \times 3} \times \left(\frac{k}{j}\right)^3 = 20\,000$ *[1 mark]*

$j^6 \times 20 \times \left(\frac{1}{j^3}\right) \times k^3 = 20\,000$

$j^6 \times j^{-3} \times k^3 = 1000 \Rightarrow j^3 \times k^3 = 1000 \Rightarrow (jk)^3 = 1000,$

so $jk = \sqrt[3]{1000} = 10$ *[1 mark for correct rearrangement]*

b) Write an expression for the coefficient of x and then solve simultaneously with the equation $jk = 10$.

coefficient of x: $j^6 \times \dfrac{6}{1} \times \dfrac{k}{j} = 37\,500$ *[1 mark]*

$$j^6 \times j^{-1} \times k \times 6 = 37\,500 \implies kj^5 = 6250$$

From a), $jk = 10$, so $k = \dfrac{10}{j}$

$kj^5 = \dfrac{10}{j} \times j^5 = 6250$ *[1 mark for using jk = 10]*

$\implies 10 \times j^{-1} \times j^5 = 6250 \implies j^4 = 625 \implies j = \pm 5$

But j and k are positive so $j = 5$ *[1 mark]*

Now input $j = 5$ into $jk = 10$ to find: $k = 2$ *[1 mark]*

c) Coefficient of x^2: $b = 5^6 \times \dfrac{6 \times 5}{1 \times 2} \times \left(\dfrac{2}{5}\right)^2 = 37\,500$

[2 marks available — 1 mark for formula,
1 mark for correct answer]

3 a) Using the binomial expansion,

$(1 - x)^{-\frac{1}{2}} \approx 1 + \left(-\dfrac{1}{2}\right)(-x) + \dfrac{\left(-\frac{1}{2}\right) \times \left(-\frac{3}{2}\right)}{1 \times 2}(-x)^2$

$\qquad + \dfrac{\left(-\frac{1}{2}\right) \times \left(-\frac{3}{2}\right) \times \left(-\frac{5}{2}\right)}{1 \times 2 \times 3}(-x)^3$ *[1 mark]*

$= 1 + \dfrac{x}{2} + \dfrac{3}{8}x^2 + \dfrac{5}{16}x^3$.

So $A = \dfrac{1}{2}$, $B = \dfrac{3}{8}$, $C = \dfrac{5}{16}$ *[1 mark]*

b) (i) $(25 - 4x)^{-\frac{1}{2}}$

$= (25)^{-\frac{1}{2}}\left(1 - \dfrac{4}{25}x\right)^{-\frac{1}{2}} = \dfrac{1}{5}\left(1 - \dfrac{4}{25}x\right)^{-\frac{1}{2}}$ *[1 mark]*

$= \dfrac{1}{5}\left(1 + \dfrac{1}{2}\left(\dfrac{4}{25}x\right) + \dfrac{3}{8}\left(\dfrac{4}{25}x\right)^2 + \dfrac{5}{16}\left(\dfrac{4}{25}x\right)^3\right)$ *[1 mark]*

$= \dfrac{1}{5}\left(1 + \dfrac{1}{2}\left(\dfrac{4}{25}x\right) + \dfrac{3}{8}\left(\dfrac{16}{625}x^2\right) + \dfrac{5}{16}\left(\dfrac{64}{15\,625}x^3\right)\right)$

$= \dfrac{1}{5}\left(1 + \dfrac{2}{25}x + \dfrac{6}{625}x^2 + \dfrac{4}{3125}x^3\right)$ *[1 mark]*

$= \dfrac{1}{5} + \dfrac{2}{125}x + \dfrac{6}{3125}x^2 + \dfrac{4}{15\,625}x^3$ *[1 mark]*

(ii) The expansion is valid for

$\left|\dfrac{-4x}{25}\right| < 1 \implies \dfrac{|-4\,|\,|x|}{25} < 1 \implies |x| < \dfrac{25}{4}$ *[1 mark]*

c) (i) $25 - 4x = 20 \implies x = \dfrac{5}{4}$ *[1 mark]*

So $\dfrac{1}{\sqrt{20}} = \left(25 - 4\left(\dfrac{5}{4}\right)\right)^{-\frac{1}{2}}$

$\approx \dfrac{1}{5} + \dfrac{2}{125}\left(\dfrac{5}{4}\right) + \dfrac{6}{3125}\left(\dfrac{5}{4}\right)^2 + \dfrac{4}{15\,625}\left(\dfrac{5}{4}\right)^3$ *[1 mark]*

$= \dfrac{1}{5} + \dfrac{2}{125}\left(\dfrac{5}{4}\right) + \dfrac{6}{3125}\left(\dfrac{25}{16}\right) + \dfrac{4}{15\,625}\left(\dfrac{125}{64}\right)$

$= \dfrac{1}{5} + \dfrac{1}{50} + \dfrac{3}{1000} + \dfrac{1}{2000}$

$= \dfrac{447}{2000}$ *[1 mark for correct simplification]*

(ii) Percentage error

$= \left|\dfrac{\text{real value} - \text{estimate}}{\text{real value}}\right| \times 100$

$= \left|\dfrac{\dfrac{1}{\sqrt{20}} - \dfrac{447}{2000}}{\dfrac{1}{\sqrt{20}}}\right| \times 100$ *[1 mark]*

$= 0.0004776... = 0.0478\%$ (3 s.f.) *[1 mark]*

4 a) $(27 + 4x)^{\frac{1}{3}} = 27^{\frac{1}{3}}\left(1 + \dfrac{4}{27}x\right)^{\frac{1}{3}} = 3\left(1 + \dfrac{4}{27}x\right)^{\frac{1}{3}}$ *[1 mark]*

$\approx 3\left[1 + \left(\dfrac{1}{3}\right)\left(\dfrac{4}{27}x\right) + \dfrac{\frac{1}{3} \times -\frac{2}{3}}{1 \times 2}\left(\dfrac{4}{27}x\right)^2\right]$ *[1 mark]*

$= 3\left[1 + \left(\dfrac{1}{3}\right)\left(\dfrac{4}{27}x\right) + \left(-\dfrac{1}{9}\right)\left(\dfrac{16}{729}x^2\right)\right]$

$= 3\left(1 + \dfrac{4}{81}x - \dfrac{16}{6561}x^2\right)$ *[1 mark]*

$= 3 + \dfrac{4}{27}x - \dfrac{16}{2187}x^2$ *[1 mark]*

b) $27 + 4x = 26.2 \implies x = -0.2$

This lies within $|x| < \dfrac{16}{3}$, so it's a valid approximation, and x is small, so higher powers can be ignored.

$\sqrt[3]{26.2} = (26.2)^{\frac{1}{3}} \approx 3 + \dfrac{4}{27}(-0.2) - \dfrac{16}{2187}(-0.2)^2$

$= 3 - 0.0296296... - 0.0002926...$

$= 2.9700777... = 2.970078$ (6 d.p.)

[2 marks available — 1 mark for substituting x = -0.2 into the expansion from part a), 1 mark for correct answer]

5 a) (i) $\sqrt{\dfrac{1 + 3x}{1 - 5x}} = \dfrac{\sqrt{1 + 3x}}{\sqrt{1 - 5x}}$

$= (1 + 3x)^{\frac{1}{2}}(1 - 5x)^{-\frac{1}{2}}$ *[1 mark]*

$(1 + 3x)^{\frac{1}{2}} = 1 + \dfrac{1}{2}(3x) + \dfrac{\frac{1}{2} \times -\frac{1}{2}}{1 \times 2}(3x)^2 + ...$

$\approx 1 + \dfrac{3}{2}x - \dfrac{9}{8}x^2$ *[1 mark]*

$(1 - 5x)^{-\frac{1}{2}} = 1 + \left(-\dfrac{1}{2}\right)(-5x) + \dfrac{-\frac{1}{2} \times -\frac{3}{2}}{1 \times 2}(-5x)^2 + ...$

$\approx 1 + \dfrac{5}{2}x + \dfrac{75}{8}x^2$ *[1 mark]*

$\sqrt{\dfrac{1 + 3x}{1 - 5x}} \approx \left(1 + \dfrac{3}{2}x - \dfrac{9}{8}x^2\right)\left(1 + \dfrac{5}{2}x + \dfrac{75}{8}x^2\right)$ *[1 mark]*

$\approx 1 + \dfrac{5}{2}x + \dfrac{75}{8}x^2 + \dfrac{3}{2}x + \dfrac{15}{4}x^2 - \dfrac{9}{8}x^2$

(ignoring any terms in x^3 or above)

$= 1 + 4x + 12x^2$ *[1 mark for correct simplification]*

(ii) Expansion of $(1 + 3x)^{\frac{1}{2}}$ is valid for: $|3x| < 1 \implies |x| < \dfrac{1}{3}$.

Expansion of $(1 - 5x)^{-\frac{1}{2}}$ is valid for:

$|-5x| < 1 \implies |-5|\,|x| < 1 \implies |x| < \dfrac{1}{5}$.

The combined expansion is valid for the narrower of these two ranges, so the expansion of $\sqrt{\dfrac{1 + 3x}{1 - 5x}}$ is valid for: $|x| < \dfrac{1}{5}$.

[2 marks available — 1 mark for identifying the valid range of the expansion as being the narrower of the two valid ranges shown, 1 mark for correct answer]

b) $x = \dfrac{1}{15} \implies \sqrt{\dfrac{1 + 3x}{1 - 5x}} = \sqrt{\dfrac{1 + \frac{3}{15}}{1 - \frac{5}{15}}} = \sqrt{\dfrac{\left(\frac{18}{15}\right)}{\left(\frac{10}{15}\right)}} = \sqrt{\dfrac{18}{10}} = \sqrt{1.8}$

$\sqrt{1.8} \approx 1 + 4\left(\dfrac{1}{15}\right) + 12\left(\dfrac{1}{15}\right)^2 = 1 + \dfrac{4}{15} + \dfrac{4}{75} = \dfrac{33}{25}$

[2 marks available — 1 mark for substituting $x = \dfrac{1}{15}$ into the expansion from part a), 1 mark for correct simplification]

6 a) $2 - 18x \equiv A(1 - 2x)^2 + B(5 + 4x)(1 - 2x) + C(5 + 4x)$ *[1 mark]*

Let $x = \dfrac{1}{2}$, then:

$2 - 9 = 7C \implies -7 = 7C \implies C = -1$ *[1 mark]*

Let $x = -\dfrac{5}{4}$, then:

$2 - \dfrac{45}{2} = \dfrac{49}{4}A \implies \dfrac{49}{2} = \dfrac{49}{4}A \implies A = 2$ *[1 mark]*

Equating the coefficients of the x^2 terms:

$0 = 4A - 8B = 8 - 8B \implies 8B = 8 \implies B = 1$ *[1 mark]*

b) $f(x) = \dfrac{2}{(5 + 4x)} + \dfrac{1}{(1 - 2x)} - \dfrac{1}{(1 - 2x)^2}$

$= 2(5 + 4x)^{-1} + (1 - 2x)^{-1} - (1 - 2x)^{-2}$

Expand each bracket separately:

$(5 + 4x)^{-1} = 5^{-1}\left(1 + \dfrac{4}{5}x\right)^{-1} = \dfrac{1}{5}\left(1 + \dfrac{4}{5}x\right)^{-1}$

$= \dfrac{1}{5}\left(1 + (-1)\left(\dfrac{4}{5}x\right) + \dfrac{-1 \times -2}{1 \times 2}\left(\dfrac{4}{5}x\right)^2 + ...\right)$

$= \dfrac{1}{5}\left(1 - \dfrac{4}{5}x + \dfrac{16}{25}x^2 + ...\right) = \dfrac{1}{5} - \dfrac{4}{25}x + \dfrac{16}{125}x^2 + ...$

$(1 - 2x)^{-1} = 1 + (-1)(-2x) + \dfrac{-1 \times -2}{1 \times 2}(-2x)^2 + ...$

$= 1 + 2x + 4x^2 + ...$

$(1 - 2x)^{-2} = 1 + (-2)(-2x) + \dfrac{-2 \times -3}{1 \times 2}(-2x)^2 + ...$

$= 1 + 4x + 12x^2 + ...$

Putting it all together gives (ignoring any terms in x^3 or above):

$f(x) \approx 2\left(\frac{1}{5} - \frac{4}{25}x + \frac{16}{125}x^2\right) + (1 + 2x + 4x^2) - (1 + 4x + 12x^2)$

$= \frac{2}{5} - \frac{8}{25}x + \frac{32}{125}x^2 + 1 + 2x + 4x^2 - 1 - 4x - 12x^2$

$= \frac{2}{5} - \frac{58}{25}x - \frac{968}{125}x^2$

[7 marks available — 1 mark for rewriting f(x) in the form $A(5 + 4x)^{-1} + B(1 - 2x)^{-1} + C(1 - 2x)^{-2}$, 1 mark for taking out a factor of 5 from $(5 + 4x)^{-1}$, 1 mark for correct expansion of $(5 + 4x)^{-1}$, 1 mark for correct expansion of $(1 - 2x)^{-1}$, 1 mark for correct expansion of $(1 - 2x)^{-2}$, 1 mark for correct constant and x-terms in final answer, 1 mark for correct x^2-term in final answer]

c) Expansion of $(5 + 4x)^{-1}$ is valid for

$\left|\frac{4x}{5}\right| < 1 \Rightarrow \frac{4|x|}{5} < 1 \Rightarrow |x| < \frac{5}{4}$

Expansions of $(1 - 2x)^{-1}$ and $(1 - 2x)^{-2}$ are valid for

$\left|\frac{-2x}{1}\right| < 1 \Rightarrow \frac{2|x|}{1} < 1 \Rightarrow |x| < \frac{1}{2}$

The combined expansion is valid for the narrower of these two ranges. So the expansion of f(x) is valid for $|x| < \frac{1}{2}$.

Claire had the wrong inequality sign in the second inequality OR she incorrectly combined the two inequalities instead of using the narrower of the two of the ranges. *[2 marks available — 1 mark for explaining the error Claire had made, 1 mark for correct range]*

Pages 30-36: Trigonometry

1 To find the area of the sector, you need the angle in radians:

$(120° \div 180°) \times \pi = \frac{2\pi}{3}$ radians *[1 mark]*

Now use the arc length to find r:

Arc length $S = r\theta$, so

$40 = \frac{2\pi}{3} \times r$ *[1 mark]*

$r = 40 \div \frac{2\pi}{3} = \frac{60}{\pi}$ *[1 mark]*

Finally, use this value of r to find the area:

$A = \frac{1}{2}r^2\theta = \frac{1}{2} \times (\frac{60}{\pi})^2 \times \frac{2\pi}{3}$ *[1 mark]*

$= \frac{1200}{\pi} = 381.9718... = 382$ cm² (to the nearest cm²) *[1 mark]*

2 a) (i) In the diagram, x is the adjacent side of a right-angled triangle with an angle θ and hypotenuse r, so use the cos formula:

$\cos\theta = \frac{\text{adjacent}}{\text{hypotenuse}} = \frac{x}{r}$, so $x = r\cos\theta$ *[1 mark]*

(ii) As the stage is symmetrical, you know that distance y is the same on both triangles.

$\sin\theta = \frac{\text{opposite}}{\text{hypotenuse}} = \frac{y}{r}$, so $y = r\sin\theta$ *[1 mark]*

b) The total length of the bottom and straight sides is $q + q + 2r$. The top length is $2x$, so using the expression found in a), you can write this as $2r\cos\theta$.

For the curved lengths, the shaded areas are sectors of circles, and the formula for the length of one arc is given by $r\theta$.

Now add them all up to get the total perimeter:

$q + q + 2r + 2r\cos\theta + r\theta + r\theta = 2[q + r(1 + \theta + \cos\theta)]$.

Break the area down into a rectangle, a triangle and two sectors:

Area of rectangle = width × height = $2qr$

Area of triangle = $\frac{1}{2}(2r\cos\theta)(r\sin\theta) = r^2\cos\theta\sin\theta$

Area of one shaded sector = $\frac{1}{2}r^2\theta$

So the total area $A = 2qr + r^2\cos\theta\sin\theta + r^2\theta$

$= 2qr + r^2(\cos\theta\sin\theta + \theta)$.

[4 marks available — 1 mark for all individual lengths correct, 1 mark for all individual areas correct, 1 mark for each correct expression]

You could've used $\frac{1}{2} AB \sin C$ to find the area of the triangle, but then you'd need to use one of the expressions for x or y from part a) to get the final answer.

c) Substitute the given values of P and θ into the expression for the perimeter: $P = 2[q + r(1 + \theta + \cos\theta)]$

$\Rightarrow 40 = 2\left[q + r\left(1 + \frac{\pi}{3} + \cos\frac{\pi}{3}\right)\right]$

$\Rightarrow 20 = 20 = q + r\left(\frac{3}{2} + \frac{\pi}{3}\right)$ *[1 mark]*

And then into the expression for the area:

$A = 2qr + r^2(\cos\theta\sin\theta + \theta)$

$\Rightarrow A = 2qr + r^2\left(\cos\frac{\pi}{3}\sin\frac{\pi}{3} + \frac{\pi}{3}\right)$

$= 2qr + r^2\left(\frac{\sqrt{3}}{4} + \frac{\pi}{3}\right)$ *[1 mark]*

To rearrange this formula for area into the form shown in the question, you need to get rid of q. Rearrange the perimeter formula to get an expression for q in terms of r, then substitute that into the area expression:

$q = 20 - r\left(\frac{3}{2} + \frac{\pi}{3}\right)$

$A = 2qr + r^2\left(\frac{\sqrt{3}}{4} + \frac{\pi}{3}\right)$

$= 2r\left[20 - r\left(\frac{3}{2} + \frac{\pi}{3}\right)\right] + r^2\left(\frac{\sqrt{3}}{4} + \frac{\pi}{3}\right)$ *[1 mark]*

$= 40r - 2r^2\left(\frac{3}{2} + \frac{\pi}{3}\right) + r^2\left(\frac{\sqrt{3}}{4} + \frac{\pi}{3}\right)$

$= 40r - r^2\left[\left(3 + \frac{2\pi}{3}\right) - \left(\frac{\sqrt{3}}{4} + \frac{\pi}{3}\right)\right]$

$= 40 - r^2\left(3 - \frac{\sqrt{3}}{4} + \frac{\pi}{3}\right)$

So $A = 40r - kr^2$, where $k = 3 - \frac{\sqrt{3}}{4} + \frac{\pi}{3}$, as required *[1 mark]*

3 The graph of $y = \sin x$ is mapped onto the graph of $y = \sin\frac{x}{2}$ via a stretch parallel to the x-axis of scale factor 2.
The graph should appear as follows:

[3 marks available — 1 mark for sin x correct, 1 mark for sin $\frac{x}{2}$ correct, 1 mark for correct axis labelling]

4 a) Use the cosine rule:

e.g. $\cos A = \frac{50^2 + 70^2 - 90^2}{2 \times 50 \times 70}$ *[1 mark]*

$\cos A = -0.1$ *[1 mark]*

$A = 95.739...°$ *[1 mark]*

Now use this value of A to find the area:

Area $= \frac{1}{2} \times 50 \times 70 \times \sin 95.739...°$ *[1 mark]*

$= 1741.228... = 1741$ m² (nearest m²) *[1 mark]*

If you've allocated your values of a, b, c etc. differently, or found a different angle, then the numbers in your working will be different.

b) E.g. the model is unlikely to give an area accurate to the nearest square metre as the given side lengths are most likely rounded, at least to the nearest metre, possibly to the nearest 5 m or 10 m. This means that there is a large range of possible areas. / The sides are unlikely to be perfectly straight, so the model will not be accurate *[1 mark for a sensible comment]*.

5 $7 - 3\cos x = 9\sin^2 x$, and $\sin^2 x \equiv 1 - \cos^2 x$

$\Rightarrow 7 - 3\cos x = 9(1 - \cos^2 x)$

$\Rightarrow 7 - 3\cos x = 9 - 9\cos^2 x$

$\Rightarrow 9\cos^2 x - 3\cos x - 2 = 0$

Substitute y for $\cos x$ and solve $9y^2 - 3y - 2 = 0$ by factorising:

$(3y - 2)(3y + 1) = 0 \Rightarrow y = \frac{2}{3}$ or $y = -\frac{1}{3}$

So $\cos x = \frac{2}{3}$ or $\cos x = -\frac{1}{3}$

For $\cos x = \frac{2}{3}$, $x = 48.189...° = 48.2°$ (1 d.p.).

For $\cos x = -\frac{1}{3}$, $x = 109.471...° = 109.5°$ (1 d.p).

[5 marks available — 1 mark for correct substitution using trig identity, 1 mark for forming a quadratic in cos x, 1 mark for finding correct values of cos x, 1 mark for each of the 2 correct solutions]

6 a) Substituting $t = 35.26...°$ into both sides of the equation gives:
LHS: $\sin(2 \times 35.26...°) = 0.94$ (2 s.f)
RHS: $\sqrt{2}\cos(2 \times 35.26...°) = 0.47$ (2 s.f.)
$0.94 \neq 0.47$, so Adam's solution is incorrect *[1 mark]*.

b) Adam has incorrectly divided by 2:
$\tan 2t = \sqrt{2} \not\Rightarrow \tan t = \frac{\sqrt{2}}{2}$ *[1 mark]*

c) $t = -27.36...°$ is not a solution of the original equation *[1 mark]*. The error appeared because Bethan has squared the equation and then taken roots *[1 mark]*.

7 a) Use the trig identity $\tan\theta \equiv \frac{\sin\theta}{\cos\theta}$:
$\tan^2\theta + \frac{\tan\theta}{\cos\theta} = 1 \rightarrow \frac{\sin^2\theta}{\cos^2\theta} + \frac{\sin\theta}{\cos^2\theta} = 1$ *[1 mark]*
Put over a common denominator: $\frac{\sin^2\theta + \sin\theta}{\cos^2\theta} = 1$
$\Rightarrow \sin^2\theta + \sin\theta = \cos^2\theta$
Now use the identity $\cos^2\theta \equiv 1 - \sin^2\theta$ to give:
$\sin^2\theta + \sin\theta = 1 - \sin^2\theta$ *[1 mark]*
$\Rightarrow 2\sin^2\theta + \sin\theta - 1 = 0$ *[1 mark for rearrangement]*.

b) Factorising the quadratic from a) gives:
$(2\sin\theta - 1)(\sin\theta + 1) = 0$ *[1 mark]*
$\Rightarrow \sin\theta = \frac{1}{2}$ or $\sin\theta = -1$ *[1 mark]*
$\sin\theta = \frac{1}{2} \Rightarrow \theta = \frac{\pi}{6}$ and $\theta = (\pi - \frac{\pi}{6}) = \frac{5\pi}{6}$ *[1 mark for both]*
$\sin\theta = -1 \Rightarrow \theta = \frac{3\pi}{2}$, but $f(\theta)$ is not defined for $\theta = \frac{3\pi}{2}$,
so ignore this solution.
So the solutions are $\theta = \frac{\pi}{6}$ and $\frac{5\pi}{6}$ *[1 mark]*

8 a) The start and end points of the cos curve (with restricted domain) are $(0, 1)$ and $(\pi, -1)$, so the coordinates of the start point of arccos (point A) are $(-1, \pi)$ *[1 mark]* and the coordinates of the end point (point B) are $(1, 0)$ *[1 mark]*.

b) $y = \arccos x \Rightarrow y = \cos^{-1}x \Rightarrow x = \cos y$ *[1 mark]*

c) $\arccos x = 2 \Rightarrow x = \cos 2$ *[1 mark]* $\Rightarrow x = -0.416$ *[1 mark]*

9 a) $\csc\theta = \frac{5}{3} \Rightarrow \sin\theta = \frac{3}{5}$. Solving for θ gives $\theta = 0.64350...$,
$\theta = \pi - 0.64350... = 2.49809...$
So $\theta = 0.644, 2.50$ (both to 3 s.f.)
[1 mark for each correct answer].
Sketch the graph of y = sin x or use a CAST diagram to help you find the second solution.

b) (i) The identity $\csc^2\theta \equiv 1 + \cot^2\theta$ rearranges to give $\csc^2\theta - 1 \equiv \cot^2\theta$. Putting this into the equation:
$3\csc\theta = (\csc^2\theta - 1) - 17$
$18 + 3\csc\theta - \csc^2\theta = 0$ as required
[2 marks available — 1 mark for using correct identity, 1 mark for rearranging into required form]

(ii) To factorise the expression above, let $x = \csc\theta$.
Then $18 + 3x - x^2 = 0$, so $(6 - x)(3 + x) = 0$ *[1 mark]*.
The roots of this quadratic occur at $x = 6$ and $x = -3$,
so $\csc\theta = 6$ and $\csc\theta = -3$ *[1 mark]*.
$\csc\theta = \frac{1}{\sin\theta}$, so $\sin\theta = \frac{1}{6}$ and $\sin\theta = -\frac{1}{3}$ *[1 mark]*.
$\sin\theta = \frac{1}{6} \Rightarrow \theta = \sin^{-1}\frac{1}{6} = 0.16744...$ or
$\theta = \pi - 0.16744... = 2.97414...$
$\sin\theta = -\frac{1}{3} \Rightarrow \theta = \sin^{-1}(-\frac{1}{3}) = -0.33983...$
but this is outside the required range.
So $\theta = 2\pi + (-0.33983...) = 5.94334...$ or
$\theta = \pi - (-0.33983...) = 3.48142...$

So $\theta = 0.167, 2.97$ *[1 mark]* and
$\theta = 3.48, 5.94$ *[1 mark]* (all to 3 s.f.).
You don't have to use x = cosec θ — it's just a little easier to factorise without all those pesky cosecs flying around.

10 Using the small angle approximations, $\sin\theta \approx \theta$, $\cos\theta \approx 1 - \frac{1}{2}\theta^2$ and $\tan\theta \approx \theta$. Substituting these values into the expression gives:
$4\sin\theta\tan\theta + 2\cos\theta \approx 4(\theta \times \theta) + 2(1 - \frac{1}{2}\theta^2)$
$= 4\theta^2 + 2 - \theta^2 = 2 + 3\theta^2$ as required,
where $p = 2$ and $q = 3$.
[3 marks available — 1 mark for the correct approximations, 1 mark for substituting into the expression, 1 mark for rearranging to obtain the correct answer]

11 a) $\sec x = \frac{1}{\cos x}$, so as $\cos x = \frac{8}{9}$, $\sec x = \frac{9}{8}$ *[1 mark]*.

b) The right-angled triangle with angle x, hypotenuse of length 9 and the adjacent side of length 8 (which gives the cos x value as stated) has the opposite side of length $\sqrt{9^2 - 8^2} = \sqrt{81 - 64} = \sqrt{17}$ *[1 mark]*.
So the value of $\sin x = \frac{\sqrt{17}}{9}$ (opposite / hypotenuse).
$\csc x = \frac{1}{\sin x}$, so $\csc x = \frac{9}{\sqrt{17}} = \frac{9\sqrt{17}}{17}$ *[1 mark]*.

c) For the triangle described in part b), the value of
$\tan x$ is given by opposite / adjacent $= \frac{\sqrt{17}}{8}$ *[1 mark]*.
So $\tan^2 x = \left(\frac{\sqrt{17}}{8}\right)^2 = \frac{17}{64}$ *[1 mark]*.
You could have used $\sec^2 x \equiv 1 + \tan^2 x$ here instead.

d) $\cos 2x = 2\cos^2 x - 1$. Using the known value of $\cos x$,
$\cos 2x = 2\left(\frac{8}{9}\right)^2 - 1 = 2\left(\frac{64}{81}\right) - 1 = \frac{47}{81}$
[3 marks available — 1 mark for formula for cos 2x, 1 mark for working and 1 mark for correct answer]
You could have used the other versions of the cos 2x formula here ($\cos^2 x - \sin^2 x$ or $1 - 2\sin^2 x$) — just use the value you found for sin x in part b).

12 a) $\frac{1 + \cos x}{2} = \frac{1}{2}\left(1 + \cos 2\left(\frac{x}{2}\right)\right)$
$= \frac{1}{2}\left(1 + \left(2\cos^2\frac{x}{2} - 1\right)\right)$
$= \frac{1}{2}\left(2\cos^2\frac{x}{2}\right) = \cos^2\frac{x}{2}$
[2 marks available – 1 mark for using the correct identity, 1 mark for the correct rearrangement]

b) As $\cos^2\frac{x}{2} = 0.75$, $\frac{1 + \cos x}{2} = 0.75$.
So $1 + \cos x = 1.5$
$\cos x = 0.5$ *[1 mark]* $\Rightarrow x = \frac{\pi}{3}, \frac{5\pi}{3}$ *[1 mark]*
You should know the solutions to cos x = 0.5 — it's one of the common angles.

13 $\sin 2\theta \equiv 2\sin\theta\cos\theta$, so $3\sin 2\theta\tan\theta \equiv 6\sin\theta\cos\theta\tan\theta$ *[1 mark]*.
As $\tan\theta \equiv \frac{\sin\theta}{\cos\theta}$,
$6\sin\theta\cos\theta\tan\theta \equiv 6\sin\theta\cos\theta\frac{\sin\theta}{\cos\theta} \equiv 6\sin^2\theta$ *[1 mark]*
so $3\sin 2\theta\tan\theta = 5 \Rightarrow 6\sin^2\theta = 5$ *[1 mark]*
Then $\sin^2\theta = \frac{5}{6} \Rightarrow \sin\theta = \pm\sqrt{\frac{5}{6}} = \pm 0.9128...$ *[1 mark]*
Solving this for θ gives $\theta = 1.15, 1.99, 4.29, 5.13$ *[2 marks for all 4 correct answers, 1 mark for 2 correct answers]*
Don't forget the solutions for the negative square root as well — they're easy to miss. Drawing a sketch here is really useful — you can see that there are 4 solutions you need to find:

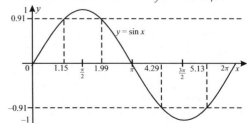

14 a) Use the double angle formula: $\cos 2\theta \equiv 1 - 2\sin^2\theta$
to replace $\cos 2\theta$:
$DE^2 = 4 - 4(1 - 2\sin^2\theta)$ *[1 mark]*
$DE^2 = 4 - 4 + 8\sin^2\theta \Rightarrow DE^2 = 8\sin^2\theta$
$DE = \sqrt{8}\sin\theta = 2\sqrt{2}\sin\theta$ *[1 mark for correct rearrangement]*

b) $P = 2DE + 2DG$
To find DG, use triangle BDG:

$\cos\theta = \dfrac{DG}{\sqrt{2}}$, so $DG = \sqrt{2}\cos\theta$ *[1 mark]*
So $P = 2(2\sqrt{2}\sin\theta) + 2(\sqrt{2}\cos\theta)$
$= 4\sqrt{2}\sin\theta + 2\sqrt{2}\cos\theta$ *[1 mark for*
correct substitution and rearrangement]

c) $4\sqrt{2}\sin\theta + 2\sqrt{2}\cos\theta \equiv R\sin(\theta + \alpha)$
$\Rightarrow 4\sqrt{2}\sin\theta + 2\sqrt{2}\cos\theta \equiv R\sin\theta\cos\alpha + R\cos\theta\sin\alpha$
$\Rightarrow$ ① $R\cos\alpha = 4\sqrt{2}$ and ② $R\sin\alpha = 2\sqrt{2}$ *[1 mark]*
② ÷ ① gives $\tan\alpha = \dfrac{1}{2} \Rightarrow \alpha = 0.464$ (3 s.f.) *[1 mark]*
①² + ②² gives:
$R^2\cos^2\alpha + R^2\sin^2\alpha = (4\sqrt{2})^2 + (2\sqrt{2})^2 = 40$
$\Rightarrow R = \sqrt{40} = 2\sqrt{10}$ *[1 mark]*
So $4\sqrt{2}\sin\theta + 2\sqrt{2}\cos\theta \equiv 2\sqrt{10}\sin(\theta + 0.464)$

15 a) $\sqrt{2}\cos\theta - 3\sin\theta \equiv R\cos(\theta + \alpha)$. Using the cos addition rule,
$R\cos(\theta + \alpha) \equiv R\cos\theta\cos\alpha - R\sin\theta\sin\alpha$,
so ① $R\cos\alpha = \sqrt{2}$ and ② $R\sin\alpha = 3$ *[1 mark]*.
② ÷ ① gives $\tan a = \dfrac{3}{\sqrt{2}} \Rightarrow \alpha = 1.13$ (3 s.f.) *[1 mark]*
①² + ②² gives:
$R^2\cos^2\alpha + R^2\sin^2\alpha = (\sqrt{2})^2 + 3^2 = 11 \Rightarrow R = \sqrt{11}$ *[1 mark]*,
so $\sqrt{2}\cos\theta - 3\sin\theta = \sqrt{11}\cos(\theta + 1.13)$.

b) The equation you're trying to solve is $\sqrt{2}\cos\theta - 3\sin\theta = 3$
in the range $0 \le \theta \le 6$, so, from part a), $\sqrt{11}\cos(\theta + 1.13) = 3$.
So $\cos(\theta + 1.13) = \dfrac{3}{\sqrt{11}}$.
Solving this gives $\theta + 1.13 = 0.4405...$ *[1 mark]*.
The range of solutions becomes $1.13 \le \theta + 1.13 \le 7.13$.
To find the other values of θ within the new range,
$2\pi - 0.4405... = 5.842..., 2\pi + 0.4405... = 6.723...$ *[1 mark*
for both]. Subtracting 1.13 gives $\theta = 4.712..., 5.593...$ mins
[1 mark for both]. So the water reaches 3 feet to the right of
the sprinkler at 4 minutes 43 seconds and at
5 minutes 36 seconds *[1 mark for both]*.
You can sketch the graph to help you find all the values of θ.

c) Using part a), $d = (\sqrt{2}\cos\theta - 3\sin\theta)^4 = (\sqrt{11}\cos(\theta + 1.13))^4$
The maximum distances left and right occur at the minimum
and maximum value of d (the minimum value corresponds
to the maximum distance left). The maximum and minimum
points of $\cos(\theta + 1.13)$ are ± 1, so the maximum and minimum
values of the function inside the brackets are $\pm\sqrt{11}$.
This bracket is raised to the power 4, so the maximum distance
right is: $(\pm\sqrt{11})^4 = 121$ feet *[1 mark]*.
Since $(\sqrt{11}\cos(\theta + 1.13))^4 \ge 0$, the maximum distance left
is 0 feet *[1 mark]*.
If you didn't realise that $\left(\sqrt{11}\cos(\theta + 1.13)\right)^4$ is never negative,
you'd have got the distance left wrong.

d) E.g. the sprinkler could be positioned against a wall, so it can
never spray to the left *[1 mark for a sensible comment]*.

16 $\csc 2A \equiv \dfrac{1}{\sin 2A} \equiv \dfrac{1}{2\sin A\cos A}$

so $2\tan A\csc 2A \equiv \dfrac{2\tan A}{2\sin A\cos A} \equiv \dfrac{2\dfrac{\sin A}{\cos A}}{2\sin A\cos A}$

$\equiv \dfrac{\sin A}{\sin A\cos^2 A} \equiv \dfrac{1}{\cos^2 A}$

$\equiv \sec^2 A \equiv 1 + \tan^2 A$

[3 marks available — 1 mark for using the double angle formula
to expand $\sin 2A$, 1 mark for rearranging and simplifying with the
use of $\tan A \equiv \dfrac{\sin A}{\cos A}$, 1 mark for using $\sec^2 A \equiv 1 + \tan^2 A$ to get
into the required form]

Pages 37-40: Exponentials and Logarithms

1 Rewrite all terms as powers of p and use the laws of logs to
simplify:
$\log_p(p^4) + \log_p(p^{\frac{1}{2}}) - \log_p(p^{-\frac{1}{2}})$ *[1 mark]*
$= 4\log_p p + \dfrac{1}{2}\log_p p - \left(-\dfrac{1}{2}\right)\log_p p$ *[1 mark]*
$= 4 + \dfrac{1}{2} - \left(-\dfrac{1}{2}\right) = 4 + 1 = 5$ (as $\log_p p = 1$) *[1 mark]*

2 $5^{(z^2 - 9)} = 2^{(z - 3)}$, so taking logs of both sides gives:
$(z^2 - 9)\ln 5 = (z - 3)\ln 2$ *[1 mark]*
$\Rightarrow (z + 3)(z - 3)\ln 5 - (z - 3)\ln 2 = 0$ *[1 mark]*
$\Rightarrow (z - 3)[(z + 3)\ln 5 - \ln 2] = 0$ *[1 mark]*
$\Rightarrow z - 3 = 0$ or $(z + 3)\ln 5 - \ln 2 = 0$
$\Rightarrow z = 3$ *[1 mark]* or $z = \dfrac{\ln 2}{\ln 5} - 3$
$\Rightarrow z = 3$ or $z = -2.57$ (3 s.f.) *[1 mark]*

3 $3^{2x} = (3^x)^2$ (from the power laws), so let $y = 3^x$, then $y^2 = 3^{2x}$.
This gives a quadratic in y: $y^2 - 9y + 14 = 0$
$(y - 2)(y - 7) = 0$ *[1 mark]*, so $y = 2$ or $y = 7$
$\Rightarrow 3^x = 2$ or $3^x = 7$ *[1 mark for both]*
To solve these equations, take logs of both sides *[1 mark]*.
$3^x = 2 \Rightarrow \log 3^x = \log 2 \Rightarrow x\log 3 = \log 2$
$\Rightarrow x = \dfrac{\log 2}{\log 3} = 0.631$ (3 s.f.) *[1 mark]*
$3^x = 7 \Rightarrow \log 3^x = \log 7 \Rightarrow x\log 3 = \log 7$
$\Rightarrow x = \dfrac{\log 7}{\log 3} = 1.77$ (3 s.f.) *[1 mark]*

4 a) $y = \ln(4x - 3)$, and $x = a$ when $y = 1$.
$1 = \ln(4a - 3) \Rightarrow e^1 = 4a - 3$ *[1 mark]*
$\Rightarrow a = \dfrac{e^1 + 3}{4} = 1.43$ to 2 d.p. *[1 mark]*

b) The curve can only exist when $4x - 3 > 0$ *[1 mark]*
$\Rightarrow x > \dfrac{3}{4} \Rightarrow x > 0.75$, so $b = 0.75$ *[1 mark]*.

5 a) A is the value of y when $x = 0$, so $A = 4$ *[1 mark]*.
Now use this value to find b:
$\dfrac{4}{e} = 4e^{10b}$ *[1 mark]* $\Rightarrow 4e^{-1} = 4e^{10b}$
$\Rightarrow -1 = 10b \Rightarrow b = -0.1$ *[1 mark]*

b) The gradient of $y = Ae^{bx}$ is bAe^{bx}. Here, $A = 4$ and $b = -0.1$,
so the gradient is $-0.1 \times 4 \times e^{-0.1x} = -0.4e^{-0.1x}$ *[1 mark for a*
correct gradient expression].
Set this equal to the value given and solve:
$-0.4e^{-0.1x} = -1$ *[1 mark]* $\Rightarrow e^{-0.1x} = 2.5$
$\Rightarrow -0.1x = \ln 2.5$ *[1 mark]* $\Rightarrow x = -10\ln 2.5$ *[1 mark]*
When $x = -10\ln 2.5$, $y = 4e^{-0.1(-10\ln 2.5)} = 4e^{\ln 2.5} = 4 \times 2.5 = 10$
So the exact coordinates are $(-10\ln 2.5, 10)$ *[1 mark]*.

6 $y = e^{ax} + b$
The sketch shows that when $x = 0$, $y = -6$, so:
$-6 = e^0 + b \Rightarrow -6 = 1 + b \Rightarrow b = -7$ *[1 mark]*.
The sketch also shows that when $y = 0$, $x = \dfrac{1}{4}\ln 7$, so:
$0 = e^{(\frac{a}{4}\ln 7)} - 7$ *[1 mark]* $\Rightarrow e^{(\frac{a}{4}\ln 7)} = 7$
$\Rightarrow \dfrac{a}{4}\ln 7 = \ln 7 \Rightarrow \dfrac{a}{4} = 1 \Rightarrow a = 4$ *[1 mark]*.
The asymptote occurs as $x \to -\infty$, so $e^{4x} \to 0$,
and since $y = e^{4x} - 7$, $y \to -7$.
So the equation of the asymptote is $y = -7$ *[1 mark]*.
You could also have solved this question by thinking about the series of
transformations that would take you from the graph of $y = e^x$ to this
one.

7   The value after the first year is $0.92 \times 8000 = £7360$ *[1 mark]*.
You need to find n, the number of months after the first year
when the value falls below £4000. So solve the equation:
$7360 \times e^{\frac{n}{12}\ln\left(1-\frac{4}{100}\right)} = 4000$ *[1 mark]* $\Rightarrow e^{\frac{n}{12}\ln 0.96} = \frac{4000}{7360}$

$\Rightarrow \frac{n}{12}\ln 0.96 = \ln\left(\frac{4000}{7360}\right)$ *[1 mark]*

$\Rightarrow n = 12\ln\frac{4000}{7360} \div \ln 0.96 = 179.246$ *[1 mark]*

179 months after the first year the value is £4003.35 to the nearest
penny (i.e. > 4000), so you need to round up to 180 months.
The total number of months is: $12 + 180 = 192$ months *[1 mark]*.
*Don't forget to add the 12 at the end — that's the months from the
first year (which had a different rate of depreciation).*

8 a) You need to find t such that:
     $2100 - 1500e^{-0.15t} > 5700e^{-0.15t}$ *[1 mark]*
     $2100 > 7200e^{-0.15t}$
     $\frac{7}{24} > e^{-0.15t}$
     $\ln\frac{7}{24} > -0.15t$ *[1 mark]*
     $\ln\frac{7}{24} \div -0.15 < t$
     $t > 8.21429...$ *[1 mark]*
     So the population of Q first exceeds the population of P
     when $t > 8.21$ (3 s.f.), i.e. in the year 2018 *[1 mark]*.
     Don't forget to flip the inequality sign when you divide by −0.15.

  b)

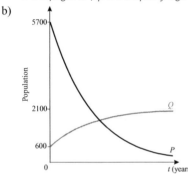

    *[2 marks available — 1 mark for correct shape of graph,
    1 mark for (0, 5700) labelled]*

  c) Bird of prey — e.g. any one of:
    - The model predicts the population of the birds of prey
      will increase, but will tend to a limit. This seems realistic,
      as the bird of prey will have to compete for the available
      sources of food as one source decreases.
    - The population grows quite slowly (especially compared to
      the rate of decrease of the other species) — this seems more
      realistic than a rapid population growth.
    - The rate of growth slows over time, which would be expected
      as food supplies dwindle.
    [1 mark for a sensible comment about the birds of prey]
    Endangered species — e.g. any one of:
    - The model predicts the population will decrease, which
      seems realistic as the birds of prey will hunt them.
    - The model predicts a very rapid decline at first, which does
      not seem realistic — you'd expect the rate of decrease to be
      slower at first.
    [1 mark for a sensible comment about the endangered species]

  d) You need to find t such that $5700e^{-0.15t} = 1000$ *[1 mark]*
    $\Rightarrow 1000 = \frac{5700}{e^{0.15t}} \Rightarrow e^{0.15t} = \frac{5700}{1000} = 5.7$.
    $\Rightarrow 0.15t = \ln 5.7 \Rightarrow t = \frac{\ln 5.7}{0.15} = 11.6031...$ years *[1 mark]*
    So the population is predicted to drop below 1000
    in the year 2021 *[1 mark]*.
    *If you set up and solved an inequality that's fine
    — you'd still get the marks.*

  e) E.g. The function could be refined so that from 2021, the
    population is predicted to stop decreasing — it could either
    level out or start increasing *[1 mark for a sensible comment]*.

9 a) $y = ab^t$, so take logs of both sides: $\log y = \log ab^t$
    Then use the laws of logs: $\log y = \log a + \log b^t$ *[1 mark]*
                         $\log y = \log a + t\log b$ *[1 mark]*
                         $\log y = t\log b + \log a$, as required.

  b) First find the values of a and b:
    Comparing $\log y = t\log b + \log a$ to $y = mx + c$ gives
    $\log b = m$, the gradient of the graph, and $\log a = c$,
    the vertical-axis intercept of the graph.
    Use points from the graph to calculate the gradient, m:
    For example, using the points $(2, 0.3)$ and $(1, 0)$:
    $m = \frac{y_2 - y_1}{x_2 - x_1} = \frac{0.3 - 0}{2 - 1} = 0.3$
    So $\log b = 0.3 \Rightarrow b = 10^{0.3}$ *[1 mark]*
    Now estimate the vertical-axis intercept to find $\log_{10} a$:
    $\log a = -0.3 \Rightarrow a = 10^{-0.3}$ *[1 mark]*.
    The equation is $y = 10^{-0.3} \times (10^{0.3})^t = 10^{0.3t - 0.3} = 10^{0.3(t-1)}$
    y is the average attendance in hundreds, so the attendance
    exceeds 5000 when $y > 50$, i.e. $10^{0.3(t-1)} > 50$ *[1 mark]*
    $0.3(t - 1) > \log 50 \Rightarrow t > \frac{\log 50}{0.3} + 1$
    $\Rightarrow t > 6.663...$ years after the 2010/2011 season *[1 mark]*.
    This is during the 2016/2017 season *[1 mark]*.
    *Don't worry if your values of a and b are slightly different
    — you should end up with the same answer though.*

  c) a is the average attendance in hundreds in the season where
    $t = 0$, i.e in the 2010/11 season *[1 mark]*.
    The attendance was around 50 supporters.

  d) For the 2024/25 season, $t = 14$, which is beyond the values of t
    given on the graph *[1 mark]*. This is extrapolation, so may not
    be accurate as the model might not hold that far in the future
    [1 mark].

Pages 41-44: Differentiation 1

1   The gradient of the tangent is the same as the gradient of the curve,
so differentiate:
$\frac{dy}{dx} = 6x^2 - 20x - 2x^{-\frac{1}{2}} = 6x^2 - 20x - \frac{2}{\sqrt{x}}$
Now put $x = 4$ into your derivative:
$6(4^2) - 20(4) - \frac{2}{\sqrt{4}} = 96 - 80 - 1 = 15$

*[4 marks available — 1 mark for differentiating, 1 mark for the
correct derivative, 1 mark for substituting in x = 4, 1 mark for the
correct answer]*

2   Differentiate $f(x)$ to find $f'(x)$:
$f'(x) = 3x^2 - 14x + 8$
So the graph of $f'(x)$ is a positive quadratic (i.e. u-shaped).
It crosses the y-axis when $x = 0$, which gives a y-value of 8.
It crosses the x-axis when $3x^2 - 14x + 8 = 0$
$\Rightarrow (3x - 2)(x - 4) = 0$, so $x = \frac{2}{3}$ and $x = 4$.
Now sketch the graph:

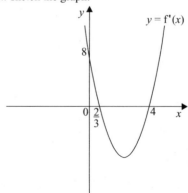

*[4 marks available — 1 mark for attempting to differentiate,
1 mark for the correct function for y = f'(x), 1 mark for the
correct shape of the graph, 1 mark for the correct x- and
y-intercepts]*

3 a) The gradient of the normal at R is the same as the gradient of the line $4y + x = 24$. Rearrange this equation to find the gradient:

$4y + x = 24 \Rightarrow 4y = 24 - x \Rightarrow y = 6 - \frac{1}{4}x$,

so gradient of normal at R $= -\frac{1}{4}$

Gradient of curve at R $= -1 \div -\frac{1}{4} = 4$

Find an expression for the gradient of the curve by differentiating $y = kx^2 - 8x - 5$:

$\frac{dy}{dx} = 2kx - 8$

At R, gradient $= 2k(2) - 8 = 4k - 8$

Put this expression equal to the value of the gradient at R to find k:

$4k - 8 = 4 \Rightarrow 4k = 12 \Rightarrow k = 3$

[5 marks available — 1 mark for finding the gradient of the normal at R, 1 mark for finding the gradient of the curve at R, 1 mark for attempting to differentiate y, 1 mark for forming an equation for k, 1 mark for the correct value of k]

   b) Gradient of tangent at R = gradient of curve = 4.

At R, $x = 2$, so $y = 3(2^2) - 8(2) - 5 = -9$ *[1 mark]*

Use these values in $y - y_1 = m(x - x_1)$ to find the equation of the tangent:

$y + 9 = 4(x - 2) \Rightarrow y + 9 = 4x - 8 \Rightarrow y = 4x - 17$ *[1 mark]*

Equate $y = 4x - 17$ and $y = 4x - \frac{1}{x^3} - 9$ to find S:

$4x - 17 = 4x - \frac{1}{x^3} - 9$ *[1 mark]*

$-8 = -\frac{1}{x^3} \Rightarrow x = \frac{1}{2}$ and $y = 4(\frac{1}{2}) - 17 = -15$

So at S, $x = \frac{1}{2}$ *[1 mark]* and $y = -15$ *[1 mark]*

4 $f'(x) = \lim_{h \to 0} \left(\frac{(8(x+h)^2 - 1) - (8x^2 - 1)}{h} \right)$ *[1 mark]*

$= \lim_{h \to 0} \left(\frac{(8(x^2 + 2xh + h^2) - 1) - (8x^2 - 1)}{h} \right)$

$= \lim_{h \to 0} \left(\frac{8x^2 + 16xh + 8h^2 - 1 - 8x^2 + 1}{h} \right)$ *[1 mark]*

$= \lim_{h \to 0} \left(\frac{16xh + 8h^2}{h} \right)$

$= \lim_{h \to 0} (16x + 8h)$ *[1 mark]*

As $h \to 0$, $16x + 8h \to 16x$, so $f'(x) = 16x$

[1 mark for letting h → 0 and obtaining the correct limit]

5 a) Differentiate $f(x)$ and set the derivative equal to zero:

$f'(x) = 8x^3 + 27$ *[1 mark]*

$8x^3 + 27 = 0$ *[1 mark]* $\Rightarrow x^3 = -\frac{27}{8}$

$\Rightarrow x = \sqrt[3]{-\frac{27}{8}} = -\frac{3}{2} = -1.5$ *[1 mark]*

When $x = -1.5$, $f(x) = 2(-1.5)^4 + 27(-1.5) = -30.375$ *[1 mark]*

So the stationary point is at $(-1.5, -30.375)$

   b) The function is increasing if the gradient is positive.

$f'(x) > 0$ if $8x^3 + 27 > 0 \Rightarrow x^3 > -\frac{27}{8} \Rightarrow x > \sqrt[3]{-\frac{27}{8}}$

$\Rightarrow x > -1.5$

The function is decreasing if the gradient is negative.

$f'(x) < 0$ if $8x^3 + 27 < 0 \Rightarrow x^3 < -\frac{27}{8} \Rightarrow x < \sqrt[3]{-\frac{27}{8}}$

$\Rightarrow x < -1.5$

[2 marks available — 1 mark for forming at least one correct inequality, 1 mark for both ranges of values correct]

   c) You know from parts a) and b) that the function has a stationary point at $(-1.5, -30.375)$ and that this is a minimum point because the function is decreasing to the left of this point and increasing to the right of it.

Find where the curve crosses the y-axis:

When $x = 0$, $f(x) = 0$, so the curve goes through the origin.

Find where the curve crosses the x-axis:

When $f(x) = 0$, $2x^4 + 27x = 0 \Rightarrow x(2x^3 + 27) = 0$

$\Rightarrow x = 0$ or $2x^3 + 27 = 0$

$\Rightarrow x^3 = -\frac{27}{2} \Rightarrow x = \sqrt[3]{-\frac{27}{2}} = -2.381$ (3 d.p.),

so the curve crosses the x-axis at $x = 0$ and $x = -2.381$.

Now use the information you've found to sketch the curve:

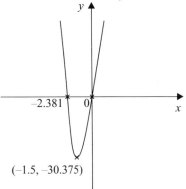

[3 marks available — 1 mark for a curve with the correct shape, 1 mark for the correct minimum point, 1 mark for the correct intercepts]

6 The graph is concave for $\frac{d^2y}{dx^2} < 0$.

$y = x^4 + 3x^3 - 6x^2 \Rightarrow \frac{dy}{dx} = 4x^3 + 9x^2 - 12x$ *[1 mark]*

$\Rightarrow \frac{d^2y}{dx^2} = 12x^2 + 18x - 12$ *[1 mark]*

$12x^2 + 18x - 12 < 0$ *[1 mark]* $\Rightarrow 2x^2 + 3x - 2 < 0$

$\Rightarrow (2x - 1)(x + 2) < 0$

$(2x - 1)(x + 2) < 0$ for $x < \frac{1}{2}$ and $x > -2$,

so the graph is concave for $-2 < x < \frac{1}{2}$ *[1 mark]*.

To check your inequality think about the shape of the graph of $\frac{d^2y}{dx^2}$ — it's a positive quadratic, so is less than O between −2 and $\frac{1}{2}$.

7 a) At a point of inflection, $f''(x) = 0$, so find $f''(x)$.

$f(x) = 3x^3 + 9x^2 + 25x \Rightarrow f'(x) = 9x^2 + 18x + 25$ *[1 mark]*

$\Rightarrow f''(x) = 18x + 18$ *[1 mark]*

$f''(-1) = 18(-1) + 18 = 0$, so $f''(x) = 0$ at $x = -1$ *[1 mark]*

To confirm that this is a point of inflection, you need to check what's happening either side of $x = -1$:

For $x < -1$, $f''(x) < 0$ and for $x > -1$, $f''(x) > 0$ *[1 mark]*.

The curve changes from concave to convex, so $x = -1$ is a point of inflection *[1 mark]*.

   b) At a stationary point, $f'(x) = 0$.

$f'(x) = 9x^2 + 18x + 25$

$f'(-1) = 9(-1)^2 + 18(-1) + 25 = 16$. Since $16 \neq 0$, the point of inflection is not a stationary point. *[2 marks available — 1 mark for finding f '(−1), 1 mark for a correct explanation of why this isn't a stationary point]*

   c) $f(x)$ is an increasing function for all values of x if $f'(x) > 0$ for all x.

From a), $f'(x) = 9x^2 + 18x + 25$

Complete the square to show that $f'(x) > 0$:

$f'(x) = 9x^2 + 18x + 25$

$\Rightarrow f'(x) = 9(x^2 + 2x) + 25$

$\Rightarrow f'(x) = 9(x + 1)^2 - 9 + 25$

$\Rightarrow f'(x) = 9(x + 1)^2 + 16$ *[1 mark]*

$(x + 1)^2 \geq 0$, so $f'(x)$ has a minimum value of 16

So $f'(x) > 0$ for all x, which means that $f(x)$ is an increasing function for all values of x *[1 mark]*.

8 a) Surface area of the container = sum of the areas of all 5 faces $= x^2 + x^2 + xy + xy + xy = 2x^2 + 3xy$ *[1 mark]*

40 litres $= 40\,000$ cm^3

Volume of the container = length × width × height

$= x^2y = 40\,000$ cm^3 *[1 mark]* $\Rightarrow y = \frac{40\,000}{x^2}$

Put this into the formula for the area:

$A = 2x^2 + 3xy = 2x^2 + 3x\left(\frac{40\,000}{x^2}\right)$ *[1 mark]*

$= 2x^2 + \frac{120\,000}{x}$ *[1 mark]*

138

b) To find stationary points, first find $\frac{dA}{dx}$:

$\frac{dA}{dx} = 4x - \frac{120\,000}{x^2}$ *[1 mark for attempting to differentiate, 1 mark for the correct function]*

Then find the value of x where $\frac{dA}{dx} = 0$:

$4x - \frac{120\,000}{x^2} = 0$ *[1 mark]* $\Rightarrow x^3 = 30\,000$

$\Rightarrow x = 31.07... = 31.1$ cm (3 s.f.) *[1 mark]*

To check if it's a minimum, find $\frac{d^2A}{dx^2}$:

$\frac{d^2A}{dx^2} = 4 + \frac{240\,000}{x^3} = 12$ at $x = 31.07...$ *[1 mark]*.

The second derivative is positive, so it's a minimum *[1 mark]*

c) Put the value of x found in part b) into the formula for the area given in part a):

$A = 2(31.07...)^2 + \frac{120\,000}{31.07...}$ *[1 mark]* $= 5792.936...$

$= 5790$ cm^2 (3 s.f.) *[1 mark]*

d) E.g. the model does not take into account the thickness of the steel, so the minimum area needed is likely to be slightly greater than this to create the required capacity *[1 mark for a sensible comment]*.

Pages 45-50: Differentiation 2

1 a) $y = \frac{1}{\sqrt{2x - x^2}} = (2x - x^2)^{-\frac{1}{2}}$

Let $u = 2x - x^2$, so $y = u^{-\frac{1}{2}}$ *[1 mark]*,

then $\frac{du}{dx} = 2 - 2x$ and $\frac{dy}{du} = -\frac{1}{2}u^{-\frac{3}{2}}$

Using the chain rule: $\frac{dy}{dx} = \frac{dy}{du} \times \frac{du}{dx} = -\frac{1}{2}u^{-\frac{3}{2}} \times (2 - 2x)$

$= -\frac{1}{2}(2x - x^2)^{-\frac{3}{2}} \times (2 - 2x)$ *[1 mark]*

$= -\frac{2 - 2x}{2(\sqrt{2x - x^2})^3} = \frac{x - 1}{(\sqrt{2x - x^2})^3}$ *[1 mark]*

So at $(1, 1)$, $\frac{dy}{dx} = 0$ *[1 mark]*.

b) $x = (4y + 10)^3$. Let $u = 4y + 10$, so $x = u^3$ *[1 mark]*,

then $\frac{du}{dy} = 4$ and $\frac{dx}{du} = 3u^2$

Using the chain rule: $\frac{dx}{dy} = \frac{dx}{du} \times \frac{du}{dy} = 3u^2 \times 4$

$= 12(4y + 10)^2$ *[1 mark]*

So $\frac{dy}{dx} = \frac{1}{\left(\frac{dx}{dy}\right)} = \frac{1}{12(4y + 10)^2}$ *[1 mark]*

So at $(8, -2)$, $\frac{dy}{dx} = \frac{1}{48}$ *[1 mark]*

You could have found the answer by rearranging the equation to get y on its own and then differentiating normally.

2 a) Replace h with x in the height formula:

$x = \sqrt{\frac{2}{3}}a \Rightarrow a = \sqrt{\frac{3}{2}}x$ *[1 mark]*

Now substitute $a = \sqrt{\frac{3}{2}}x$ into the expression for volume:

$V = \frac{\sqrt{2}}{12}\left(\sqrt{\frac{3}{2}}x\right)^3 = \frac{\sqrt{2}}{12} \times \frac{3\sqrt{3}}{2\sqrt{2}}x^3 = \frac{\sqrt{3}}{8}x^3$

[1 mark for substitution and correct simplification]

b) From the question, you know that the rate of change of volume with respect to time, $\frac{dV}{dt}$, is 240. And differentiating the expression for volume from part a) with respect to x gives

$\frac{dV}{dx} = \frac{3\sqrt{3}x^2}{8}$ *[1 mark]*

Using the chain rule: $\frac{dx}{dt} = \frac{dx}{dV} \times \frac{dV}{dt}$ *[1 mark]*

$= \frac{1}{\left(\frac{dV}{dx}\right)} \times \frac{dV}{dt} = \frac{8}{3\sqrt{3}x^2} \times 240 = \frac{640}{\sqrt{3}x^2}$ *[1 mark]*

So when $x = 8$, $\frac{dx}{dt} = \frac{640}{64\sqrt{3}} = \frac{10}{\sqrt{3}} = \frac{10\sqrt{3}}{3}$ cm min^{-1}

[1 mark for substitution of x = 8, 1 mark for correct answer in surd form]

c) $\frac{dV}{dt} = \frac{dV}{dx} \times \frac{dx}{dt}$ *[1 mark]*

$= \frac{3\sqrt{3}x^2}{8} \times \frac{32}{9\sqrt{3}} = \frac{4x^2}{3}$ *[1 mark]*

So when $x = 12$, $\frac{dV}{dt} = \frac{4 \times 144}{3} = 192$ cm^3 min^{-1} *[1 mark]*

3 a) $y = e^{2x} - 5e^x + 3x$, so using chain rule:

$\frac{dy}{dx} = 2e^{2x} - 5e^x + 3$

[1 mark for $2e^{2x}$, 1 mark for the other two terms correct]

b) Stationary points occur when $\frac{dy}{dx} = 0$, so:

$2e^{2x} - 5e^x + 3 = 0$ *[1 mark]*.

This looks like a quadratic, so substitute $z = e^x$ and factorise:

$2z^2 - 5z + 3 = 0 \Rightarrow (2z - 3)(z - 1) = 0$ *[1 mark]*

So the solutions are:

$2z - 3 = 0 \Rightarrow z = \frac{3}{2} \Rightarrow e^x = \frac{3}{2} \Rightarrow x = \ln\frac{3}{2}$ *[1 mark]*, and

$z - 1 = 0 \Rightarrow z = 1 \Rightarrow e^x = 1 \Rightarrow x = \ln 1 = 0$ *[1 mark]*.

c) To determine the nature of the stationary points,

find $\frac{d^2y}{dx^2}$ at $x = 0$ and $x = \ln\frac{3}{2}$:

$\frac{d^2y}{dx^2} = 4e^{2x} - 5e^x$ *[1 mark]*

When $x = 0$, $\frac{d^2y}{dx^2} = 4e^0 - 5e^0 = 4 - 5 = -1$,

so $\frac{d^2y}{dx^2} < 0$, which means the point is a maximum *[1 mark]*.

When $x = \ln\frac{3}{2}$, $\frac{d^2y}{dx^2} = 4e^{2\ln\frac{3}{2}} - 5e^{\ln\frac{3}{2}} = 4\left(\frac{3}{2}\right)^2 - 5\left(\frac{3}{2}\right) = \frac{3}{2}$,

so $\frac{d^2y}{dx^2} > 0$, which means the point is a minimum *[1 mark]*.

4 a) Use the chain rule to find $\frac{dy}{dx}$:

$y = 2^{3x}$, so $u = 3x$ and $y = 2^u$, *[1 mark]*

so $\frac{du}{dx} = 3$ and $\frac{dy}{du} = 2^u \ln 2$

$\Rightarrow \frac{dy}{dx} = 2^u \ln 2 \times 3 = 3(2^{3x} \ln 2)$ *[1 mark]*

At $x = 1$, $\frac{dy}{dx} = 3(2^3 \ln 2) = 24 \ln 2$ *[1 mark]*

b) y is an increasing function for all values of x

if $\frac{dy}{dx} > 0$ for all x *[1 mark]*. From a), $\frac{dy}{dx} = 3(2^{3x} \ln 2)$.

Since $2^{3x} > 0$ for all x and 3 and $\ln 2$ are both positive,

$3(2^{3x} \ln 2) > 0$ for all x *[1 mark]*.

5 $f(x) = \cos x$, so:

$f'\left(\frac{\pi}{3}\right) = \lim_{h \to 0}\left(\frac{\cos\left(\frac{\pi}{3} + h\right) - \cos\left(\frac{\pi}{3}\right)}{h}\right)$ *[1 mark]*

$= \lim_{h \to 0}\left(\frac{\cos\frac{\pi}{3}\cos h - \sin\frac{\pi}{3}\sin h - \cos\left(\frac{\pi}{3}\right)}{h}\right)$ *[1 mark]*

$= \lim_{h \to 0}\left(\frac{\cos\frac{\pi}{3}(\cos h - 1) - \sin\frac{\pi}{3}\sin h}{h}\right)$

$= \lim_{h \to 0}\left(\frac{\frac{1}{2}\left(-\frac{1}{2}h^2\right) - \frac{\sqrt{3}}{2}h}{h}\right)$

[1 mark for using small angle approximations for each of cos h and sin h]

$= \lim_{h \to 0}\left(-\frac{h}{4} - \frac{\sqrt{3}}{2}\right)$ *[1 mark]*

As $h \to 0$, $-\frac{h}{4} \to 0$, so $f'\left(\frac{\pi}{3}\right) \to -\frac{\sqrt{3}}{2}$ *[1 mark]*

6 $y = \ln x (5x - 2)^3$, so use the product rule

with $u = \ln x$ and $v = (5x - 2)^3$.

First use the chain rule to find $\frac{dv}{dx} = 15(5x - 2)^2$.

So $u = \ln x$, $\frac{du}{dx} = \frac{1}{x}$, $v = (5x - 2)^3$, $\frac{dv}{dx} = 15(5x - 2)^2$

Then $\frac{dy}{dx} = u\frac{dv}{dx} + v\frac{du}{dx} = \ln x \times 15(5x - 2)^2 + (5x - 2)^3 \times \frac{1}{x}$

$= 15 \ln x (5x - 2)^2 + \frac{(5x - 2)^3}{x}$

$= (5x - 2)^2\left[15 \ln x + \frac{(5x - 2)}{x}\right]$

$= (5x - 2)^2\left[15 \ln x + 5 - \frac{2}{x}\right]$

So $a = 15$, $b = 5$ and $c = -2$.

Answers

[4 marks available — 1 mark each for finding expressions for du/dx and dv/dx, 1 mark for putting these expressions into the product rule, 1 mark for the answer in the correct form with correct a, b and c]

7   Use the quotient rule to find $\frac{dy}{dx}$:

$u = 5 - 2x \Rightarrow \frac{du}{dx} = -2$

$v = 3x^2 + 3x \Rightarrow \frac{dv}{dx} = 6x + 3$

$\frac{dy}{dx} = \frac{v\frac{du}{dx} - u\frac{dv}{dx}}{v^2}$

$= \frac{(3x^2 + 3x)(-2) - (5 - 2x)(6x + 3)}{(3x^2 + 3x)^2}$

$= \frac{-6x^2 - 6x - (30x + 15 - 12x^2 - 6x)}{(3x^2 + 3x)^2} = \frac{6x^2 - 30x - 15}{(3x^2 + 3x)^2}$

At a stationary point, $\frac{dy}{dx} = 0$.

$\frac{6x^2 - 30x - 15}{(3x^2 + 3x)^2} = 0 \Rightarrow 6x^2 - 30x - 15 = 0 \Rightarrow 2x^2 - 10x - 5 = 0$

[6 marks available — 1 mark each for du/dx and dv/dx, 1 mark for putting these expressions into the quotient rule to find dy/dx, 1 mark for correct expression for dy/dx, 1 mark for setting dy/dx equal to zero, 1 mark for simplifying to show that $2x^2 - 10x - 5 = 0$]

8   $y = 4x^2 \ln x$, so use the product rule to find $\frac{dy}{dx}$:

$u = 4x^2$ and $v = \ln x$, so $\frac{du}{dx} = 8x$ and $\frac{dv}{dx} = \frac{1}{x}$ *[1 mark]*

So $\frac{dy}{dx} = 4x^2 \times \frac{1}{x} + \ln x \times 8x$

$= 4x + 8x \ln x$ *[1 mark]*

Now use the product rule with $u = 8x$ and $v = \ln x$ to find $\frac{d^2y}{dx^2}$:

$u = 8x$ and $v = \ln x$, so $\frac{du}{dx} = 8$ and $\frac{dv}{dx} = \frac{1}{x}$ *[1 mark]*

So $\frac{d^2y}{dx^2} = 4 + \left(8x \times \frac{1}{x} + \ln x \times 8\right)$

$= 4 + 8 + 8 \ln x$

$= 12 + 8 \ln x$ *[1 mark]*

The curve is concave for $\frac{d^2y}{dx^2} < 0$.

$12 + 8 \ln x < 0$ *[1 mark]* $\Rightarrow \ln x < -1.5 \Rightarrow x < e^{-1.5}$

You know that $x > 0$, so the curve is concave for $0 < x < e^{-1.5}$ *[1 mark]*.

The curve is convex for $\frac{d^2y}{dx^2} > 0$.

$12 + 8 \ln x > 0$ *[1 mark]* $\Rightarrow \ln x > -1.5 \Rightarrow x > e^{-1.5}$, so the curve is convex for $x > e^{-1.5}$ *[1 mark]*.

You could have factorised $\frac{dy}{dx}$ to get $4x(1 + 2 \ln x)$. This would make the following few steps a bit different, but the answer will be the same.

9   a) $y = \frac{4x - 1}{\tan x}$, so use the quotient rule:

$u = 4x - 1 \Rightarrow \frac{du}{dx} = 4$

$v = \tan x \Rightarrow \frac{dv}{dx} = \sec^2 x$

$\frac{dy}{dx} = \frac{v\frac{du}{dx} - u\frac{dv}{dx}}{v^2} = \frac{4\tan x - (4x - 1)\sec^2 x}{\tan^2 x}$

$= \frac{4}{\tan x} - \frac{(4x - 1)\sec^2 x}{\tan^2 x}$

Since $\frac{1}{\tan x} = \cot x$, $\sec^2 x = \frac{1}{\cos^2 x}$ and $\tan^2 x = \frac{\sin^2 x}{\cos^2 x}$:

$\frac{dy}{dx} = 4 \cot x - \frac{(4x - 1)}{\cos^2 x \left(\frac{\sin^2 x}{\cos^2 x}\right)} = 4 \cot x - \frac{(4x - 1)}{\sin^2 x}$

Since $\frac{1}{\sin^2 x} = \text{cosec}^2 x$:

$\frac{dy}{dx} = 4 \cot x - (4x - 1)\, \text{cosec}^2 x$

[3 marks available — 1 mark for correct expressions for du/dx and dv/dx, 1 mark for correct use of the quotient rule, and 1 mark for reaching the correct expression for dy/dx]

Alternatively, you could have used the product rule with $y = (4x - 1)\cot x$.

b)   Maximum point is when $\frac{dy}{dx} = 0$:

$4 \cot x - (4x - 1)\, \text{cosec}^2 x = 0$ *[1 mark]*

Dividing through by $\text{cosec}^2 x$ gives

$\frac{4 \cot x}{\text{cosec}^2 x} - (4x - 1) = 0$ *[1 mark]*

$\frac{4 \cot x}{\text{cosec}^2 x} = \frac{4 \cos x \sin^2 x}{\sin x} = 4 \cos x \sin x$ *[1 mark]*,

and using the double angle formula, $\sin 2x = 2 \sin x \cos x$, so $4 \cos x \sin x = 2 \sin 2x$.

So $2 \sin 2x - 4x + 1 = 0$ *[1 mark]*.

You're told that the point is a maximum, so you don't need to differentiate again to check.

10   $\frac{dx}{dt} = 2t$, $\frac{dy}{dt} = 3t^2 + 2$

So $\frac{dy}{dx} = \frac{dy}{dt} \div \frac{dx}{dt} = \frac{3t^2 + 2}{2t}$

Substitute $t = 1$ into this expression to find the gradient of the curve at $t = 1$:

$\frac{dy}{dx} = \frac{3(1)^2 + 2}{2(1)} = \frac{5}{2}$

Now find the coordinates of the point where $t = 1$:

$x = 1^2 + 1 = 2$ and $y = 1^3 + 2(1) = 3$

Now use the gradient at $t = 1$ and the point $(2, 3)$ to find the equation of the tangent:

$y - 3 = \frac{5}{2}(x - 2) \Rightarrow y - 3 = \frac{5}{2}x - 5 \Rightarrow y = \frac{5}{2}x - 2$

[5 marks available — 1 mark for a correct method to find dy/dx, 1 mark for a correct expression for dy/dx, 1 mark for substituting $t = 1$ to find the gradient, 1 mark for finding the coordinates when $t = 1$, 1 mark for the correct equation]

11   $y = \cos^{-1} x \Rightarrow \cos y = x$ *[1 mark]*

Differentiate with respect to x:

$-\sin y \times \frac{dy}{dx} = 1 \Rightarrow \frac{dy}{dx} = -\frac{1}{\sin y}$ *[1 mark]*

Using the identity $\cos^2 y + \sin^2 y = 1$,

$\sin^2 y = 1 - \cos^2 y \Rightarrow \sin y = \sqrt{1 - \cos^2 y}$

So $\frac{dy}{dx} = -\frac{1}{\sqrt{1 - \cos^2 y}}$ *[1 mark]*

As $\cos y = x$, this expression becomes $\frac{dy}{dx} = -\frac{1}{\sqrt{1 - x^2}}$ *[1 mark]*

You could have differentiated x with respect to y then used $\frac{dy}{dx} = \frac{1}{\left(\frac{dx}{dy}\right)}$ instead of using implicit differentiation here.

12   a) $\frac{dx}{d\theta} = \frac{\cos \theta}{2}$, $\frac{dy}{d\theta} = 2 \sin 2\theta$

So $\frac{dy}{dx} = \frac{dy}{d\theta} \div \frac{dx}{d\theta}$

$= 2 \sin 2\theta \div \frac{\cos \theta}{2} = \frac{4 \sin 2\theta}{\cos \theta}$

[2 marks available — 1 mark for a correct method to find dy/dx, 1 mark for a correct expression for dy/dx]

b)   When $\theta = \frac{\pi}{6}$, $\frac{dy}{dx} = \frac{4 \sin\left(\frac{\pi}{3}\right)}{\cos\left(\frac{\pi}{6}\right)} = \frac{4\left(\frac{\sqrt{3}}{2}\right)}{\left(\frac{\sqrt{3}}{2}\right)} = 4$ *[1 mark]*

$x = \frac{1}{2}\sin\frac{\pi}{6} - 3 = \frac{1}{2} \times \frac{1}{2} - 3 = \frac{1}{4} - 3 = -\frac{11}{4}$

$y = 5 - \cos\frac{\pi}{3} = 5 - \frac{1}{2} = \frac{9}{2}$

[1 mark for x and y values both correct]

So using $y - y_1 = m(x - x_1)$:

$y - \frac{9}{2} = 4\left(x + \frac{11}{4}\right) \Rightarrow 2y - 9 = 8x + 22$

So the equation in the correct form is $8x - 2y + 31 = 0$ *[1 mark]*

c)   Substitute $x = \frac{\sin \theta}{2} - 3$ and $y = 5 - \cos 2\theta$ into $y = -8x - 20$:

$5 - \cos 2\theta = -8\left(\frac{\sin \theta}{2} - 3\right) - 20$ *[1 mark]*

$\Rightarrow 1 - \cos 2\theta + 4 \sin \theta = 0$

Using the double angle identity, $\cos 2\theta \equiv 1 - 2\sin^2\theta$,
so $1 - (1 - 2\sin^2\theta) + 4\sin\theta = 0$ *[1 mark]*
$\Rightarrow 1 - 1 + 2\sin^2\theta + 4\sin\theta = 0$
$\Rightarrow 2\sin^2\theta + 4\sin\theta = 0$
$\Rightarrow \sin^2\theta + 2\sin\theta = 0$ *[1 mark]*
$\Rightarrow \sin\theta(\sin\theta + 2) = 0$
This gives $\sin\theta = 0$ or $\sin\theta = -2$ *[1 mark]*.
Since $\sin\theta$ must be between -1 and 1, $\sin\theta = -2$ is not valid.
So substitute $\sin\theta = 0$ into the expression for x:
$x = \dfrac{\sin\theta}{2} - 3 \Rightarrow x = 0 - 3 = -3$ *[1 mark]*
Substitute $x = -3$ into $y = -8x - 20$
to get $y = -8(-3) - 20 = 4$ *[1 mark]*
So P is the point $(-3, 4)$.

d) Rewrite the equation for y using the identity
$\cos 2\theta \equiv 1 - 2\sin^2\theta$:
$y = 5 - \cos 2\theta = 5 - (1 - 2\sin^2\theta) = 4 + 2\sin^2\theta$ *[1 mark]*
Now rearrange the equation for x to make $\sin\theta$ the subject:
$x = \dfrac{\sin\theta}{2} - 3 \Rightarrow 2x + 6 = \sin\theta$ *[1 mark]*
Substitute this into the equation for y:
$y = 4 + 2\sin^2\theta = 4 + 2(2x + 6)^2$
$\Rightarrow y = 4 + 2(4x^2 + 24x + 36)$
$\Rightarrow y = 8x^2 + 48x + 76$ *[1 mark]*

13 a) Use implicit differentiation to find $\dfrac{dy}{dx}$:
$\dfrac{d}{dx}x^3 + \dfrac{d}{dx}x^2y = \dfrac{d}{dx}y^2 - \dfrac{d}{dx}1$
$3x^2 + \dfrac{d}{dx}x^2y = \dfrac{d}{dx}y^2 - 0$
$3x^2 + \dfrac{d}{dx}x^2y = \dfrac{d}{dy}y^2\dfrac{dy}{dx}$
$3x^2 + \dfrac{d}{dx}x^2y = 2y\dfrac{dy}{dx}$
$3x^2 + x^2\dfrac{d}{dx}y + y\dfrac{d}{dx}x^2 = 2y\dfrac{dy}{dx}$
$3x^2 + x^2\dfrac{dy}{dx} + 2xy = 2y\dfrac{dy}{dx}$
Rearrange to make $\dfrac{dy}{dx}$ the subject:
$(2y - x^2)\dfrac{dy}{dx} = 3x^2 + 2xy$
$\dfrac{dy}{dx} = \dfrac{3x^2 + 2xy}{2y - x^2}$
[4 marks available — 1 mark for the correct differentiation of x^3 and -1, 1 mark for the correct differentiation of y^2, 1 mark for the correct differentiation of x^2y and 1 mark for rearranging to find the correct answer]

b) Substitute $x = 1$ into the original equation:
$x = 1 \Rightarrow (1)^3 + (1)^2y = y^2 - 1$ *[1 mark]*
$\Rightarrow y^2 - y - 2 = 0 \Rightarrow (y - 2)(y + 1) = 0$
$\Rightarrow y = 2$ or $y = -1$
$a > b$, so $a = 2$, $b = -1$ *[1 mark]*

c) At $Q(1, -1)$,
$\dfrac{dy}{dx} = \dfrac{3(1)^2 + 2(1)(-1)}{2(-1) - (1)^2} = \dfrac{3 - 2}{-2 - 1} = -\dfrac{1}{3}$ *[1 mark]*
So the gradient of the normal at Q is $-1 \div -\dfrac{1}{3} = 3$ *[1 mark]*
$y - y_1 = m(x - x_1) \Rightarrow y + 1 = 3(x - 1)$
$\Rightarrow y = 3x - 4$ *[1 mark]*

14 Use implicit differentiation to find $\dfrac{dy}{dx}$:
$\dfrac{d}{dx}(\sin\pi x) - \dfrac{d}{dx}\left(\cos\dfrac{\pi y}{2}\right) = \dfrac{d}{dx}(0.5)$
$\pi\cos\pi x - \dfrac{d}{dx}\left(\cos\dfrac{\pi y}{2}\right) = 0$ *[1 mark]*
$\pi\cos\pi x - \dfrac{d}{dy}\left(\cos\dfrac{\pi y}{2}\right)\dfrac{dy}{dx} = 0$
$\pi\cos\pi x + \left(\dfrac{\pi}{2}\sin\dfrac{\pi y}{2}\right)\dfrac{dy}{dx} = 0$ *[1 mark]*
Rearrange to make $\dfrac{dy}{dx}$ the subject:
$\dfrac{dy}{dx} = -\dfrac{\pi\cos\pi x}{\dfrac{\pi}{2}\sin\dfrac{\pi y}{2}} = -\dfrac{2\cos\pi x}{\sin\dfrac{\pi y}{2}}$
The stationary point is where the gradient is zero.
$\dfrac{dy}{dx} = 0 \Rightarrow -\dfrac{2\cos\pi x}{\sin\dfrac{\pi y}{2}} = 0$ *[1 mark]* $\Rightarrow \cos\pi x = 0$ *[1 mark]*
$\Rightarrow x = \dfrac{1}{2}$ or $x = \dfrac{3}{2}$ in the range $0 \le x \le 2$ *[1 mark]*

Answers

Put these values in the equation of the curve:
$x = \dfrac{3}{2} \Rightarrow \sin\dfrac{3\pi}{2} - \cos\dfrac{\pi y}{2} = 0.5$
$\Rightarrow -1 - \cos\dfrac{\pi y}{2} = 0.5$
$\Rightarrow \cos\dfrac{\pi y}{2} = -1.5$
So y has no solutions when $x = \dfrac{3}{2}$ *[1 mark]*
$x = \dfrac{1}{2} \Rightarrow \sin\dfrac{\pi}{2} - \cos\dfrac{\pi y}{2} = 0.5$
$\Rightarrow 1 - \cos\dfrac{\pi y}{2} = 0.5$
$\Rightarrow \cos\dfrac{\pi y}{2} = 0.5$
$\Rightarrow \dfrac{\pi y}{2} = \dfrac{\pi}{3}$
$\Rightarrow y = \dfrac{2}{3}$ in the range $0 \le y \le 2$ *[1 mark]*
So the stationary point of the graph of $\sin\pi x - \cos\dfrac{\pi y}{2} = 0.5$
for the given ranges of x and y is at $\left(\dfrac{1}{2}, \dfrac{2}{3}\right)$.

Pages 51-55: Integration 1

1 $\int\left(\dfrac{x^2 + 3}{\sqrt{x}}\right)dx = \int(x^{\frac{3}{2}} + 3x^{-\frac{1}{2}})dx = \dfrac{x^{\frac{5}{2}}}{(5/2)} + \dfrac{3x^{\frac{1}{2}}}{(1/2)} + C$
$= \dfrac{2}{5}x^{\frac{5}{2}} + 6x^{\frac{1}{2}} + C = \dfrac{2}{5}\sqrt{x^5} + 6\sqrt{x} + C$
[3 marks available — 1 mark for writing both terms as powers of x, 1 mark for increasing the power of one term by 1, 1 mark for the correct integrated terms and adding C]

2 To find f(x), integrate f$'(x)$:
f$(x) = \int\left(2x + 5\sqrt{x} + \dfrac{6}{x^2}\right)dx = \int(2x + 5x^{\frac{1}{2}} + 6x^{-2})dx$
$= \dfrac{2x^2}{2} + 5\left(\dfrac{x^{\frac{3}{2}}}{(\frac{3}{2})}\right) + \left(\dfrac{6x^{-1}}{-1}\right) + C$
f$(x) = x^2 + \dfrac{10\sqrt{x^3}}{3} - \dfrac{6}{x} + C$
[4 marks available for the above working — 1 mark for writing all terms as powers of x, 1 mark for increasing the power of one term by 1, 1 mark for two correct simplified terms, 1 mark for the third correct integrated term and adding C]
You've been given a point on the curve so calculate the value of C:
If $y = 7$ when $x = 3$, then
$3^2 + \dfrac{10\sqrt{3^3}}{3} - \dfrac{6}{3} + C = 7$ *[1 mark]*
$9 + 10\sqrt{3} - 2 + C = 7$
$7 + 10\sqrt{3} + C = 7 \Rightarrow C = -10\sqrt{3}$
f$(x) = x^2 + \dfrac{10\sqrt{x^3}}{3} - \dfrac{6}{x} - 10\sqrt{3}$ *[1 mark]*

3 To find the area of region A, you need to integrate the function between $x = 2$ and $x = 4$:
Area $= \int_2^4 \dfrac{2}{\sqrt{x^3}}dx = \int_2^4 2x^{-\frac{3}{2}}dx = \left[-2(2x^{-\frac{1}{2}})\right]_2^4 = \left[\dfrac{-4}{x^{\frac{1}{2}}}\right]_2^4 = \left[\dfrac{-4}{\sqrt{x}}\right]_2^4$
$= \left(\dfrac{-4}{\sqrt{4}}\right) - \left(\dfrac{-4}{\sqrt{2}}\right) = \dfrac{-4}{2} + \dfrac{4}{\sqrt{2}} = -2 + \dfrac{4\sqrt{2}}{2}$
$= 2\sqrt{2} - 2$ as required
[5 marks available — 1 mark for writing down the correct integral to find, 1 mark for integrating correctly, 1 mark for correct handling of the limits, 1 mark for rationalising the denominator, 1 mark for rearranging to give the answer in the correct form]

4 $\int_p^{4p}\left(\dfrac{1}{\sqrt{x}} - 4x^3\right)dx = \int_p^{4p}(x^{-\frac{1}{2}} - 4x^3)dx = \left[\dfrac{x^{\frac{1}{2}}}{\frac{1}{2}} - \dfrac{4x^4}{4}\right]_p^{4p}$
$= \left[2x^{\frac{1}{2}} - x^4\right]_p^{4p} = \left[2\sqrt{x} - x^4\right]_p^{4p}$
$= (2\sqrt{4p} - (4p)^4) - (2\sqrt{p} - p^4)$
$= (4\sqrt{p} - 256p^4) - (2\sqrt{p} - p^4)$
$= 2\sqrt{p} - 255p^4$
[4 marks available — 1 mark for increasing the power of one term by 1, 1 mark for the correct integrated terms, 1 mark for correct handling of the limits, 1 mark for simplifying to get the final answer]

5   To find the shaded area, you need to integrate the function between –1 and 0.5 and add it to the integral of the function between 0.5 and 2 (making this value positive first).

$$\int_{-1}^{0.5} (2x^3 - 3x^2 - 11x + 6)\,dx = \left[\frac{2x^4}{4} - \frac{3x^3}{3} - \frac{11x^2}{2} + 6x\right]_{-1}^{0.5}$$

$$= \left[\frac{x^4}{2} - x^3 - \frac{11}{2}x^2 + 6x\right]_{-1}^{0.5}$$

$$= \left(\frac{(0.5)^4}{2} - (0.5)^3 - \frac{11}{2}(0.5)^2 + 6(0.5)\right)$$

$$- \left(\frac{(-1)^4}{2} - (-1)^3 - \frac{11}{2}(-1)^2 + 6(-1)\right)$$

$$= 1.53125 - (-10) = 11.53125$$

So the area between –1 and 0.5 is 11.53125.

$$\int_{0.5}^{2} (2x^3 - 3x^2 - 11x + 6)\,dx = \left[\frac{x^4}{2} - x^3 - \frac{11}{2}x^2 + 6x\right]_{0.5}^{2}$$

$$= \left(\frac{(2)^4}{2} - (2)^3 - \frac{11}{2}(2)^2 + 6(2)\right) - 1.53125$$

$$= -10 - 1.53125 = -11.53125$$

So the area between 0.5 and 2 is 11.53125.
So area = 11.53125 + 11.53125 = 23.0625

[6 marks available — 1 mark for considering the area above and below the x-axes separately, 1 mark for increasing the power of one term by 1, 1 mark for the correct integral, 1 mark for finding the area between –1 and 0.5, 1 mark for finding the area between 0.5 and 2, 1 mark for adding the areas to get the correct answer]

If you'd just integrated between –1 and 2, you'd have ended up with an answer of 0, as the areas cancel each other out.

6   Evaluate the integral, treating k as a constant:

$$\int_{\sqrt{2}}^{2} (8x^3 - 2kx)\,dx = \left[\frac{8x^4}{4} - \frac{2kx^2}{2}\right]_{\sqrt{2}}^{2} = \left[2x^4 - kx^2\right]_{\sqrt{2}}^{2}$$

$$= (2(2)^4 - k(2)^2) - (2(\sqrt{2})^4 - k(\sqrt{2})^2)$$

$$= (32 - 4k) - (8 - 2k) = 24 - 2k$$

You know that the value of this integral is $2k^2$, so set this expression equal to $2k^2$ and solve to find k:

$24 - 2k = 2k^2$

$0 = 2k^2 + 2k - 24 \Rightarrow k^2 + k - 12 = 0 \Rightarrow (k + 4)(k - 3) = 0$

So $k = -4$ or $k = 3$

[5 marks available — 1 mark for increasing the power of one term by 1, 1 mark for the correct integrated terms, 1 mark for substituting in the limits, 1 mark for setting this expression equal to $2k^2$, 1 mark for solving the quadratic to find both values of k]

7   a) A and B are the points where the two lines intersect, so

$$\frac{8}{x^2} = 9 - x^2 \;\textit{[1 mark]} \Rightarrow 8 = 9x^2 - x^4$$

$$\Rightarrow x^4 - 9x^2 + 8 = 0$$

$$\Rightarrow (x^2 - 8)(x^2 - 1) = 0 \;\textit{[1 mark]}$$

$x^2 = 8 \Rightarrow x = \pm\sqrt{8} = \pm 2\sqrt{2} = 2\sqrt{2}$ as $x \geq 0$

$x^2 = 1 \Rightarrow x = \pm 1 = 1$ as $x \geq 0$ *[1 mark for both values of x]*

When $x = 1$, $y = 8$, and when $x = 2\sqrt{2}$, $y = 1$

So $A = (1, 8)$ *[1 mark]* and $B = (2\sqrt{2}, 1)$ *[1 mark]*.

b)   The shaded region is the area under $y = 9 - x^2$ minus the area under $y = \frac{8}{x^2}$ from $x = 1$ to $x = 2\sqrt{2}$, so integrate:

$$\int_{1}^{2\sqrt{2}} \left(9 - x^2 - \frac{8}{x^2}\right) dx = \left[9x - \frac{x^3}{3} + \frac{8}{x}\right]_{1}^{2\sqrt{2}}$$

$$= \left(18\sqrt{2} - \frac{(2\sqrt{2})^3}{3} + \frac{8}{2\sqrt{2}}\right) - \left(9 - \frac{1}{3} + 8\right)$$

$$= 18\sqrt{2} - \frac{16\sqrt{2}}{3} + 2\sqrt{2} - 9 + \frac{1}{3} - 8$$

$$= \frac{44}{3}\sqrt{2} - \frac{50}{3}$$

[4 marks available — 1 mark for increasing the power of one term by 1, 1 mark for the correct integral, 1 mark for correct handling of the limits, 1 mark for simplifying to get the final answer]

8   First find the equation of line N — it's a normal to the curve, so differentiate: $\frac{dy}{dx} = \frac{1}{2}x^{-\frac{1}{2}} - x$

[1 mark for differentiating, 1 mark for correct derivative]

When $x = 1$, $\frac{dy}{dx} = \frac{1}{2} - 1 = -\frac{1}{2}$ *[1 mark]*, so the gradient of the normal to the curve at this point is $-1 \div -\frac{1}{2} = 2$ *[1 mark]*.

So the equation of the normal is:

$y - \frac{3}{2} = 2(x - 1) \Rightarrow y = 2x - \frac{1}{2}$ *[1 mark]*.

So the area you need is:

$$\int_{0}^{1} \left[\left(\sqrt{x} - \frac{1}{2}x^2 + 1\right) - \left(2x - \frac{1}{2}\right)\right] dx \;\textit{[1 mark]}$$

$$= \int_{0}^{1} \left(x^{\frac{1}{2}} - \frac{1}{2}x^2 - 2x + \frac{3}{2}\right) dx$$

$$= \left[\frac{2}{3}x^{\frac{3}{2}} - \frac{1}{6}x^3 - x^2 + \frac{3}{2}x\right]_{0}^{1} \;\textit{[1 mark]}$$

$$= \left(\frac{2}{3} - \frac{1}{6} - 1 + \frac{3}{2}\right) - 0 = 1$$

[1 mark for substituting in the limits correctly, 1 mark for the correct answer]

You could have done this one by working out each bit separately.

9   a)   At $x = 0$: $0 = 8t^3 \Rightarrow t = 0$

At $x = 1$: $1 = 8t^3 \Rightarrow t^3 = \frac{1}{8} \Rightarrow t = \frac{1}{2}$

[1 mark for both correct]

b)   From the chain rule, $\int y\,dx = \int y\frac{dx}{dt}\,dt$.

Find $\frac{dx}{dt}$: $x = 8t^3$, so $\frac{dx}{dt} = 24t^2$ *[1 mark]*

Using the t-limits 0 and $\frac{1}{2}$ from a), the area of A is:

$$\int_{0}^{\frac{1}{2}} (2t - 16t^4)(24t^2)\,dt = \int_{0}^{\frac{1}{2}} (48t^3 - 384t^6)\,dt \;\textit{[1 mark]}$$

$$= \left[12t^4 - \frac{384}{7}t^7\right]_{0}^{\frac{1}{2}} \;\textit{[1 mark]}$$

$$= \frac{3}{4} - \frac{3}{7} - 0 = \frac{9}{28} \;\textit{[1 mark]}$$

For parametric integration, you have to use the t-limits — otherwise you'll get it wrong. You might have been tempted to convert the parametric equations into a Cartesian equation to integrate — this method is usually <u>much harder</u>.

10   Find the value of the integral in terms of p:

$$\int_{2p}^{6p} \frac{x^3 + 4x^2}{x^3}\,dx = \int_{2p}^{6p} \left(1 + \frac{4}{x}\right) dx = [x + 4\ln x]_{2p}^{6p}$$

$$= (6p + 4\ln 6p) - (2p + 4\ln 2p)$$

$$= 4p + 4\ln\frac{6p}{2p} = 4p + 4\ln 3$$

You know the value of the integral, so set these expressions equal to each other and solve for p:

$4p + 4\ln 3 = 4\ln 12$

$p = \ln 12 - \ln 3 = \ln\frac{12}{3} = \ln 4$

[4 marks available — 1 mark for simplifying the fraction and integrating, 1 mark for the correct integral, 1 mark for substituting in the limits correctly and simplifying, 1 mark for setting the expressions equal to each other and solving to find p]

11   First, rewrite as partial fractions:

$$\frac{1}{x(3x - 2)} \equiv \frac{A}{x} + \frac{B}{3x - 2} \Rightarrow 1 \equiv A(3x - 2) + Bx \;\textit{[1 mark]}$$

Substituting $x = 0$ gives $A = -\frac{1}{2}$

Substituting $x = \frac{2}{3}$ gives $B = \frac{3}{2}$ *[1 mark for both]*

$$\Rightarrow \frac{1}{x(3x - 2)} \equiv \frac{3}{2(3x - 2)} - \frac{1}{2x}$$

So $\int \frac{1}{x(3x - 2)}\,dx = \int \left(\frac{3}{2(3x - 2)} - \frac{1}{2x}\right) dx$ *[1 mark]*

$$= \int \left(\frac{3}{2}(3x - 2)^{-1} - \frac{1}{2}x^{-1}\right) dx$$

$$= \frac{1}{2}\ln|3x - 2| - \frac{1}{2}\ln|x| + C$$

$$= \frac{1}{2}\ln\left|\frac{3x - 2}{x}\right| + C$$

[2 marks for the correct answer — deduct 1 mark if the constant is missing]

You could also write C as ln k, making your final answer $\frac{1}{2}\ln\left|\frac{k(3x-2)}{x}\right|$. As long as you include the constant, then you'll get all the marks.

12 a) $\dfrac{3x+5}{(3x+1)(1-x)} \equiv \dfrac{A}{3x+1} + \dfrac{B}{1-x}$ *[1 mark]*

$\Rightarrow 3x+5 \equiv A(1-x) + B(3x+1)$

Substituting $x = 1$ gives $B = 2$

Substituting $x = -\dfrac{1}{3}$ gives $A = 3$ *[1 mark for both]*

$\Rightarrow \dfrac{3x+5}{(3x+1)(1-x)} \equiv \dfrac{3}{(3x+1)} + \dfrac{2}{(1-x)}$ *[1 mark]*

b) $\displaystyle\int_{-2}^{4} \dfrac{3x+5}{(3x+1)(1-x)}\, dx = \int_{-2}^{4}\left(\dfrac{3}{(3x+1)} + \dfrac{2}{(1-x)}\right) dx$

$= [\ln|3x+1| - 2\ln|1-x|]_{-2}^{4}$

$= (\ln 13 - 2\ln|-3|) - (\ln|-5| - 2\ln 3)$

$= \ln 13 - 2\ln 3 - \ln 5 + 2\ln 3$

$= \ln 13 - \ln 5 = \ln \dfrac{13}{5}$

[3 marks available — 1 mark for the correct integral, 1 mark for substituting in the limits, 1 mark for simplifying to give the answer in the correct form]

13 $4x^2 + 4x - 3 = (2x+3)(2x-1)$, so $\dfrac{4x-10}{4x^2+4x-3} \equiv \dfrac{4x-10}{(2x+3)(2x-1)}$

Rewrite this as partial fractions:

$\dfrac{4x-10}{(2x+3)(2x-1)} \equiv \dfrac{A}{2x+3} + \dfrac{B}{2x-1}$

$\Rightarrow 4x-10 \equiv A(2x-1) + B(2x+3)$

Substituting $x = \dfrac{1}{2}$ gives $B = -2$

Substituting $x = -\dfrac{3}{2}$ gives $A = 4$

$\Rightarrow \dfrac{4x-10}{(2x+3)(2x-1)} \equiv \dfrac{4}{(2x+3)} - \dfrac{2}{(2x-1)}$

Now integrate:

$\displaystyle\int_{-1}^{1} \dfrac{4x-10}{4x^2+4x-3}\, dx = \int_{-1}^{1}\left(\dfrac{4}{(2x+3)} - \dfrac{2}{(2x-1)}\right) dx$

$= [2\ln|2x+3| - \ln|2x-1|]_{-1}^{1}$

$= (2\ln 5 - \ln 1) - (2\ln 1 - \ln|-3|)$

$= 2\ln 5 + \ln 3 = \ln(5^2 \times 3) = \ln 75$

So $k = 75$.

[7 marks available — 1 mark for factorising the denominator, 1 mark for writing as partial fractions with the correct denominators, 1 mark for finding A or B, 1 mark for the correct partial fractions, 1 mark for the correct integral, 1 mark for substituting in the limits, 1 mark for simplifying to give the answer in the correct form]

Pages 56-59: Integration 2

1 $\displaystyle\int_{\frac{\pi}{12}}^{\frac{\pi}{8}} \sin 2x\, dx = \left[-\dfrac{1}{2}\cos 2x\right]_{\frac{\pi}{12}}^{\frac{\pi}{8}}$

$= \left(-\dfrac{1}{2}\cos\left(\dfrac{\pi}{4}\right)\right) - \left(-\dfrac{1}{2}\cos\left(\dfrac{\pi}{6}\right)\right)$

$= -\dfrac{1}{2\sqrt{2}} + \dfrac{\sqrt{3}}{4} = \dfrac{\sqrt{3}-\sqrt{2}}{4}$

[3 marks available — 1 mark for integrating correctly, 1 mark for substituting in the limits, 1 mark for the correct answer in surd form]

2 Use the identity $\sec^2 x \equiv 1 + \tan^2 x$ to write $3\tan^2\frac{x}{2} + 3$ as $3\sec^2\frac{x}{2}$. The integral becomes:

$\displaystyle\int\left(3\tan^2\left(\dfrac{x}{2}\right) + 3\right) dx = \int 3\sec^2\left(\dfrac{x}{2}\right) dx = 6\tan\left(\dfrac{x}{2}\right) + C$

[3 marks available — 1 mark for rewriting the integral in terms of $\sec^2\frac{x}{2}$, 1 mark for integrating correctly, 1 mark for the correct answer including + C]

3 $\dfrac{d}{dx}(e^{\tan x}) = \sec^2 x\, e^{\tan x}$, so the integral is of the form

$\displaystyle\int \dfrac{du}{dx} f'(u)\, dx = f(u) + C$, so $\int \sec^2 x\, e^{\tan x}\, dx = e^{\tan x} + C$

[2 marks available — 1 mark for integrating correctly, 1 mark for the correct answer including + C]

4 If $x = \sin\theta$, then $\dfrac{dx}{d\theta} = \cos\theta$, so $dx = d\theta\cos\theta$ *[1 mark]*.

Change the limits: as $x = \sin\theta$, $\theta = \sin^{-1} x$,

so when $x = 0$, $\theta = 0$ and when $x = \dfrac{1}{2}$, $\theta = \dfrac{\pi}{6}$ *[1 mark]*.

Putting all this into the integral gives:

$\displaystyle\int_0^{\frac{1}{2}} \dfrac{x}{1-x^2}\, dx = \int_0^{\frac{\pi}{6}} \dfrac{\sin\theta}{1-\sin^2\theta}\cos\theta\, d\theta$ *[1 mark]*

Using the identity $\sin^2\theta + \cos^2\theta \equiv 1$, replace $1 - \sin^2\theta$:

$\displaystyle\int_0^{\frac{\pi}{6}} \dfrac{\sin\theta\cos\theta}{\cos^2\theta}\, d\theta = \int_0^{\frac{\pi}{6}} \dfrac{\sin\theta}{\cos\theta}\, d\theta = \int_0^{\frac{\pi}{6}} \tan\theta\, d\theta$ *[1 mark]*

$= [-\ln|\cos\theta|]_0^{\frac{\pi}{6}}$ *[1 mark]*

$= -\ln\left|\cos\dfrac{\pi}{6}\right| + \ln|\cos 0|$ *[1 mark]*

$= -\ln\dfrac{\sqrt{3}}{2} + \ln 1 = -\ln\sqrt{3} + \ln 2$

$= \ln 2 - \ln\sqrt{3}\ \left(= \ln\dfrac{2}{\sqrt{3}}\right)$ *[1 mark]*

Be careful when the substitution is of the form $x = f(\theta)$ rather than $\theta = f(x)$ — when you change the limits, you need to find the inverse of $f(\theta)$ then put in the given values of x.

5 As $u = \ln x$, $\dfrac{du}{dx} = \dfrac{1}{x}$, so $x\, du = dx$ *[1 mark]*. The limits $x = 1$ and $x = 2$ become $u = \ln 1 = 0$ and $u = \ln 2$ *[1 mark]*.

$\left(\dfrac{\ln x}{\sqrt{x}}\right)^2 = \dfrac{(\ln x)^2}{x}$. So the integral is:

$\displaystyle\int_0^{\ln 2} \dfrac{u^2}{x} x\, du = \int_0^{\ln 2} u^2\, du$ *[1 mark]*

$= \left[\dfrac{u^3}{3}\right]_0^{\ln 2}$ *[1 mark]*

$= \dfrac{(\ln 2)^3}{3} = 0.111\ (3\ \text{s.f.})$ *[1 mark]*

6 Let $u = 4x$, so $\dfrac{du}{dx} = 4$. Let $\dfrac{dv}{dx} = e^{-2x}$, so $v = -\dfrac{1}{2}e^{-2x}$.

Putting this into the formula for integration by parts gives:

$\displaystyle\int 4xe^{-2x} dx = \left[4x\left(-\dfrac{1}{2}e^{-2x}\right)\right] - \int 4\left(-\dfrac{1}{2}e^{-2x}\right) dx$

$= -2xe^{-2x} + \int 2e^{-2x} dx$

$= -2xe^{-2x} - e^{-2x} + C\ (= -e^{-2x}(2x+1) + C)$

[4 marks available — 1 mark for correct choice of u and dv/dx, 1 mark for correct differentiation and integration to obtain du/dx and v, 1 mark for correct integration by parts method, 1 mark for answer including + C]

7 a) Let $u = x$, so $\dfrac{du}{dx} = 1$.

Let $\dfrac{dv}{dx} = \sin 4x$, so $v = -\dfrac{1}{4}\cos 4x$

Using integration by parts,

$\displaystyle\int x\sin 4x\, dx = -\dfrac{1}{4}x\cos 4x - \int -\dfrac{1}{4}\cos 4x\, dx$

$= -\dfrac{1}{4}x\cos 4x + \dfrac{1}{16}\sin 4x + C$

[4 marks available — 1 mark for correct choice of u and dv/dx, 1 mark for correct differentiation and integration to obtain du/dx and v, 1 mark for correct integration by parts method, 1 mark for correct answer including + C]

b) Let $u = x^2$, so $\dfrac{du}{dx} = 2x$.

Let $\dfrac{dv}{dx} = \cos 4x$, so $v = \dfrac{1}{4}\sin 4x$

Using integration by parts,

$\displaystyle\int x^2\cos 4x\, dx = \dfrac{1}{4}x^2\sin 4x - \int \dfrac{1}{2}x\sin 4x\, dx$

$= \dfrac{1}{4}x^2\sin 4x - \dfrac{1}{2}\left(-\dfrac{1}{4}x\cos 4x + \dfrac{1}{16}\sin 4x\right) + C$

$= \dfrac{1}{4}x^2\sin 4x + \dfrac{1}{8}x\cos 4x - \dfrac{1}{32}\sin 4x + C$

[4 marks available — 1 mark for correct choice of u and dv/dx, 1 mark for correct differentiation and integration to obtain du/dx and v, 1 mark for correct integration by parts method, 1 mark for correct answer including + C]

You've already worked out $\int x\sin 4x\, dx$ in part a), so you can use this answer in part b).

8 Let $u = \ln x$, so $\dfrac{du}{dx} = \dfrac{1}{x}$.

Let $\dfrac{dv}{dx} = \dfrac{1}{2x^2} = \dfrac{1}{2}x^{-2}$, so $v = -\dfrac{1}{2}x^{-1}$

Using integration by parts,

$\displaystyle\int_1^4 \dfrac{\ln x}{2x^2}\,dx = \left[-\dfrac{\ln x}{2x}\right]_1^4 - \int_1^4 -\dfrac{1}{2x^2}\,dx = \left[-\dfrac{\ln x}{2x}\right]_1^4 - \left[\dfrac{1}{2x}\right]_1^4$

$= \left[-\dfrac{\ln 4}{8} - -\dfrac{\ln 1}{2}\right] - \left[\dfrac{1}{8} - \dfrac{1}{2}\right] = \dfrac{3}{8} - \dfrac{\ln 4}{8}\ \left(=\dfrac{3 - \ln 4}{8}\right)$

[6 marks available — 1 mark for correct choice of u and dv/dx, 1 mark for correct differentiation and integration to obtain du/dx and v, 1 mark for correct integration by parts method, 1 mark for correct integral, 1 mark for substituting in the limits, 1 mark for answer]

9 a) $\dfrac{dN}{dt} = k\sqrt{N},\ k > 0$ *[1 mark for LHS, 1 mark for RHS]*
When $N = 36$, $\dfrac{dN}{dt} = 0.36$. Putting these values into the equation gives $0.36 = k\sqrt{36}$ *[1 mark]* $= 6k \Rightarrow k = 0.06$ (the population is increasing so ignore the negative square root).
So the differential equation is $\dfrac{dN}{dt} = 0.06\sqrt{N}$ *[1 mark]*.

b) (i) $\dfrac{dN}{dt} = \dfrac{kN}{\sqrt{t}} \Rightarrow \displaystyle\int \dfrac{1}{N}\,dN = \int \dfrac{k}{\sqrt{t}}\,dt$

$\ln|N| = 2k\sqrt{t} + C$ *[1 mark]*

$\Rightarrow N = e^{2k\sqrt{t}+C} = Ae^{2k\sqrt{t}}$, where $A = e^C$ *[1 mark]*
For the initial population, $t = 0$, so $N = 25$ when $t = 0$.
Putting these values into the equation: $25 = Ae^0 \Rightarrow 25 = A$, so the equation for N is: $N = 25e^{2k\sqrt{t}}$ *[1 mark]*.

(ii) When initial population has doubled, $N = 50$ *[1 mark]*.
Put this value and the value for k into the equation and solve for t:
$50 = 25e^{2(0.05)\sqrt{t}} \Rightarrow 2 = e^{0.1\sqrt{t}} \Rightarrow \ln 2 = 0.1\sqrt{t}$ *[1 mark]*
$10\ln 2 = \sqrt{t} \Rightarrow (10\ln 2)^2 = t \Rightarrow t = 48.045$
So it will take 48 weeks *[1 mark]* (to the nearest week) for the population to double.

10 a) First solve the differential equation to find S:
$\dfrac{dS}{dt} = k\sqrt{S} \Rightarrow \dfrac{1}{\sqrt{S}}\,dS = k\,dt$

$\Rightarrow \displaystyle\int S^{-\frac{1}{2}}\,dS = \int k\,dt$

$\Rightarrow 2S^{\frac{1}{2}} = kt + C$ *[1 mark]*

$\Rightarrow S = \left(\dfrac{1}{2}(kt + C)\right)^2 = \dfrac{1}{4}(kt + C)^2$ *[1 mark]*

At the start of the campaign, $t = 0$.
Putting $t = 0$ and $S = 81$ into the equation gives:
$81 = \dfrac{1}{4}(0 + C)^2 \Rightarrow 324 = C^2 \Rightarrow C = 18$ (C must be positive, otherwise the sales would be decreasing).
This gives the equation $S = \dfrac{1}{4}(kt + 18)^2$ *[1 mark]*.

b) When $t = 0$, $S = 81$ and $\dfrac{dS}{dt} = 18$.
Substituting this into $\dfrac{dS}{dt} = k\sqrt{S}$ gives $k = 2$.
Using $S = \dfrac{1}{4}(kt + 18)^2$ with $t = 5$ and $k = 2$ gives
$\dfrac{1}{4}((5 \times 2) + 18)^2 = 196$ kg sold.
[3 marks available — 1 mark for finding the value of k, 1 mark for substituting correct values of t and k, 1 mark for answer]

c) To find the value of t when $S = 225$, solve the equation
$225 = \dfrac{1}{4}(2t + 18)^2$ *[1 mark]*:
$225 = \dfrac{1}{4}(2t + 18)^2 \Rightarrow 900 = (2t + 18)^2$
$\Rightarrow 30 = 2t + 18 \Rightarrow 12 = 2t \Rightarrow 6 = t$
So it will be 6 days *[1 mark]* before 225 kg of cheese is sold.

11 a) (i) $\dfrac{dr}{dt} = -krt,\ k > 0$ *[1 mark for RHS, 1 mark for LHS]*

(ii) S = area of curved surface + area of circular surface
$= \dfrac{1}{2}(4\pi r^2) + \pi r^2 = 3\pi r^2$ *[1 mark]*
So $\dfrac{dS}{dr} = 6\pi r$ *[1 mark]*

$\dfrac{dS}{dt} = \dfrac{dS}{dr} \times \dfrac{dr}{dt}$ *[1 mark]*
$= 6\pi r \times -krt = -6\pi kr^2 t = -2kt(3\pi r^2) = -2ktS$
[1 mark for correct substitution and simplification to required answer]

b) (i) $\dfrac{dS}{dt} = -2ktS \Rightarrow \dfrac{dS}{S} = -2kt\,dt$

$\Rightarrow \displaystyle\int \dfrac{1}{S}\,dS = \int -2kt\,dt$

$\Rightarrow \ln S = -kt^2 + \ln A$

[1 mark for correct integration of both sides, plus a constant term]

$\Rightarrow S = e^{-kt^2 + \ln A} = Ae^{-kt^2}$ *[1 mark]*

$S = 200$ at $t = 10 \Rightarrow 200 = Ae^{-100k}$
$S = 50$ at $t = 30 \Rightarrow 50 = Ae^{-900k}$
$\Rightarrow Ae^{-100k} = 4Ae^{-900k}$ *[1 mark]*
$\Rightarrow e^{-100k} = 4e^{-900k}$
$\Rightarrow -100k = \ln 4 - 900k$
$\Rightarrow 800k = \ln 4$
$\Rightarrow k = 0.00173$ (3 s.f.) *[1 mark]*
So $200 = Ae^{-100k} = Ae^{-0.173} = 0.841A$
$\Rightarrow A = 238$ (3 s.f.) *[1 mark]*
So $S = 238e^{-0.00173t^2}$

(ii) The initial surface area is given when $t = 0$
$\Rightarrow S = 238e^0 = 238$ cm^2 (3 s.f.) *[1 mark]*

c) E.g. The differential equation for the hemisphere was calculated using an expression for its surface area. The expression for the surface area of a full sphere will be different to the expression for a hemisphere of the same radius, so this differential equation will not be appropriate *[1 mark for a sensible comment]*.

Pages 60-64: Numerical Methods

1 a) To find the inverse, let $y = f(x)$, so $y = 4(x^2 - 1)$.
Now make x the subject:
$y = 4(x^2 - 1) \Rightarrow \dfrac{y}{4} = x^2 - 1 \Rightarrow \dfrac{y}{4} + 1 = x^2$
So $x = \sqrt{\dfrac{y}{4} + 1}$ *[1 mark]* (you can ignore the negative square root, as the domain of f(x) is $x \geq 0$).
Finally, replace y with x and x with $f^{-1}(x)$:
$f^{-1}(x) = \sqrt{\dfrac{x}{4} + 1}$ *[1 mark for correct inverse]*.
$y = f^{-1}(x)$ is a reflection of $y = f(x)$ in the line $y = x$ *[1 mark]*, so the point at which the lines $y = f(x)$ and $y = f^{-1}(x)$ meet is also the point where $y = f^{-1}(x)$ meets the line $y = x$.
At this point, $x = \sqrt{\dfrac{x}{4} + 1}$, so $\sqrt{\dfrac{x}{4} + 1} - x = 0$
[1 mark for setting $f^{-1}(x)$ equal to x and rearranging].

b) Let $g(x) = \sqrt{\dfrac{x}{4} + 1} - x$
If there is a root in the interval $1 < x < 2$ then there will be a change of sign for $g(x)$ between 1 and 2:
$g(1) = \sqrt{\dfrac{1}{4} + 1} - 1 = 0.1180...$
$g(2) = \sqrt{\dfrac{2}{4} + 1} - 2 = -0.7752...$ *[1 mark for both]*
There is a change of sign and the function is continuous over this interval, so there is a root in the interval $1 < x < 2$ *[1 mark]*.

c) $x_{n+1} = \sqrt{\dfrac{x_n}{4} + 1}$, and $x_0 = 1$, so:
$x_1 = \sqrt{\dfrac{1}{4} + 1} = 1.1180...$ *[1 mark]*
$x_2 = \sqrt{\dfrac{1.1180...}{4} + 1} = 1.1311...$
$x_3 = \sqrt{\dfrac{1.1311...}{4} + 1} = 1.1326...$ *[1 mark]*
So $x = 1.13$ to 3 s.f. *[1 mark]*.

d) No. If you sketch the line $y = x$ on the graph, you can see that it does not cross the curve $y = f^{-1}(x)$ more than once, so there is only one root of the equation $\sqrt{\dfrac{x}{4} + 1} - x = 0$.
[1 mark for 'No' with suitable explanation]

2 a) When the curve and line intersect, $6^x = x + 2 \Rightarrow 6^x - x - 2 = 0$.
Let $f(x) = 6^x - x - 2$. If there is a root in the interval [0.5, 1] then there will be a change of sign for $f(x)$ between 0.5 and 1:
$f(0.5) = 6^{0.5} - 0.5 - 2 = -0.0505...$
$f(1) = 6^1 - 1 - 2 = 3$ *[1 mark for both]*
There is a change of sign and the function is continuous over this interval, so there is a root in the interval [0.5, 1] *[1 mark]*.

b) Substitute $f(x) = 6^x - x - 2$ into the Newton-Raphson formula:
$$x_{n+1} = x_n - \frac{f(x_n)}{f'(x_n)} = x_n - \frac{6^{x_n} - x_n - 2}{6^{x_n}\ln 6 - 1}$$
$$= \frac{x_n(6^{x_n}\ln 6 - 1)}{6^{x_n}\ln 6 - 1} - \frac{6^{x_n} - x_n - 2}{6^{x_n}\ln 6 - 1}$$
$$= \frac{x_n 6^{x_n}\ln 6 - x_n - 6^{x_n} + x_n + 2}{6^{x_n}\ln 6 - 1}$$
$$= \frac{6^{x_n}(x_n\ln 6 - 1) + 2}{6^{x_n}\ln 6 - 1} \text{ as required}$$
[4 marks available — 1 mark for differentiating f(x) correctly, 1 mark for substituting everything into the Newton-Raphson formula, 1 mark for putting x_n and $f(x_n)$ over a common denominator, 1 mark for factorising the numerator]

c) $x_0 = 0.5$
$$x_1 = \frac{6^{(0.5)}((0.5)\ln 6 - 1) + 2}{6^{(0.5)}\ln 6 - 1} = 0.514904...$$
$x_2 = 0.514653...$
$x_3 = 0.514653...$
x_2 and x_3 are both the same to at least 5 s.f. so the x-coordinate of P is 0.5147 (4 s.f.)
[3 marks available — 1 mark for putting $x_0 = 0.5$ into the Newton-Raphson formula from part b), 1 mark for repeated iterations until all values round to the same number to an appropriate number of significant figures, 1 mark for the correct answer]
If you put 0.5 into your calculator and press =, then input $(6^{ANS} \times (ANS \times \ln 6 - 1) + 2) \div (6^{ANS} \times \ln 6 - 1)$ and keep pressing =, you'll get the iterative sequence without having to type it in each time.

d) From above, $x = 0.5147$ to 4 s.f.
If $x = 0.5147$ to 4 s.f., the upper and lower bounds are 0.51475 and 0.51465 *[1 mark]* — any value in this range would be rounded to 0.5147. $f(x) = 6^x - x - 2$, and at point P, $f(x) = 0$.
$f(0.51475) = 0.000338...$ and $f(0.51465) = -0.0000116...$
[1 mark for both f(0.51475) positive and f(0.51465) negative].
There is a change of sign, and since $f(x)$ is continuous there must be a root in this interval *[1 mark]*.

e) E.g. If the tangent has a gradient of 0, the denominator of the fraction in the iteration formula will be 0 so the Newton-Raphson method will fail for this starting value as no value of x_1 can be found. *[1 mark for any suitable explanation]*

3 a) Substitute the values you're given for x_0, $f(x_0)$ and $f'(x_0)$ into the Newton-Raphson formula:
$$x_1 = x_0 - \frac{f(x_0)}{f'(x_0)} = -1.5 - \frac{-1.625}{9.75}$$
$$= -1.33333... = -1.333 \text{ (4 s.f.)}$$
[2 marks available — 1 mark for substituting the values into the Newton-Raphson formula correctly, 1 mark for the correct answer]

b) If $x = -1.315$ to 4 s.f., the upper and lower bounds are -1.3145 and -1.3155 *[1 mark]* — any value in this range would be rounded to -1.315. $f(x) = x^3 - x^2 + 4$.
$f(-1.3145) = 0.00075...$ and $f(-1.3155) = -0.00706...$
[1 mark for both f(-1.3145) positive and f(-1.3155) negative].
There is a change of sign, and since $f(x)$ is continuous there must be a root in this interval, so the value of b must be correct to 4 s.f. *[1 mark]*.

c) The denominator of the fraction in the Newton-Raphson formula is $f'(x) = 3x^2 - 2x$ *[1 mark]*.
When $x = \frac{2}{3}$, $3x^2 - 2x = 3(\frac{2}{3})^2 - 2(\frac{2}{3}) = 0$
— so the denominator is 0, which means the Newton-Raphson method fails as no value of x_1 can be found *[1 mark]*.

4 a) The trapezium rule is given by:
$$\int_a^b y \, dx \approx \frac{1}{2}h\{(y_0 + y_n) + 2(y_1 + y_2 + ... + y_{n-1})\}$$
where n is the number of intervals (in this case 4), and h is the width of each strip:
$$h = \frac{b-a}{n} = \frac{2-0}{4} = 0.5 \text{ [1 mark]}$$
Work out each y-value:
$x_0 = 0$ $y_0 = 2^{0^2} = 2^0 = 1$
$x_1 = 0.5$ $y_1 = 2^{0.5^2} = 2^{0.25} = 1.189$ (3 d.p.)
$x_2 = 1$ $y_2 = 2^{1^2} = 2^1 = 2$
$x_3 = 1.5$ $y_3 = 2^{1.5^2} = 2^{2.25} = 4.757$ (3 d.p.)
$x_4 = 2$ $y_4 = 2^{2^2} = 2^4 = 16$ *[1 mark for all values correct]*
Now put all these values into the formula:
$$\int_0^2 2^{x^2} dx \approx \frac{0.5}{2}\{(1 + 16) + 2(1.189 + 2 + 4.757)\} \text{ [1 mark]}$$
$$= \frac{1}{4}(17 + 15.892) = 8.22 \text{ (3 s.f.) [1 mark]}$$

b) E.g. The curve is convex, so a trapezium on each strip has a greater area than that under the curve. So the trapezium rule gives an overestimate for the area.
[1 mark for a correct answer with an explanation relating to the shape of the graph].

c) To improve the accuracy of the estimate, increase the number of intervals used in the trapezium rule *[1 mark]*.

5 a) When $x = \frac{\pi}{2}$, $y = \frac{\pi}{2}\sin\frac{\pi}{2} = \frac{\pi}{2} = 1.5708$ (5 s.f.),
and when $x = \frac{3\pi}{4}$, $y = \frac{3\pi}{4}\sin\frac{3\pi}{4} = 1.6661$ (5 s.f.)
[1 mark for both]

b) The width of each strip (h) is $\frac{\pi}{4}$, so the trapezium rule gives:
Area of $R \approx \frac{1}{2} \times \frac{\pi}{4}\{(0 + 0) + 2(0.55536 + 1.5708 + 1.6661)\}$
$$= \frac{\pi}{8}\{2(3.79226)\} = 2.97843... = 2.978 \text{ (4 s.f.)}$$
[3 marks available — 1 mark for correct value of h, 1 mark for using the formula correctly, 1 mark for correct answer]

c) Let $u = x$, so $\frac{du}{dx} = 1$. Let $\frac{dv}{dx} = \sin x$, so $v = -\cos x$
[1 mark for correct choice of u and dv/dx, 1 mark for correct du/dx and v]
Using integration by parts,
$$\int_0^\pi x\sin x \, dx = [-x\cos x]_0^\pi - \int_0^\pi -\cos x \, dx \text{ [1 mark]}$$
$$= [-x\cos x]_0^\pi + [\sin x]_0^\pi \text{ [1 mark]}$$
$$= (-\pi\cos\pi + 0\cos 0)$$
$$+ (\sin\pi - \sin 0) \text{ [1 mark]}$$
$$= (\pi - 0) + (0) = \pi \text{ [1 mark]}$$
If you'd tried to use u = sin x, you'd have ended up with a more complicated function to integrate ($x^2\cos x$).

d) To find the percentage error, divide the difference between the approximate answer and the exact answer by the exact answer and multiply by 100:
$$\left|\frac{\pi - 2.97843...}{\pi}\right| \times 100 = 5.2\% \text{ (2 s.f.)}$$
[2 marks available — 1 mark for appropriate method, 1 mark for correct answer]

6 $n = 5$, $h = \frac{4 - 1.5}{5} = 0.5$
Work out the x- and y-values (given to 5 s.f. where appropriate):
$x_0 = 1.5$ $y_0 = 2.8182$
$x_1 = 2.0$ $y_1 = 4$
$x_2 = 2.5$ $y_2 = 5.1216$
$x_3 = 3.0$ $y_3 = 6.1716$
$x_4 = 3.5$ $y_4 = 7.1364$
$x_5 = 4.0$ $y_5 = 8$
$$\int_{1.5}^4 y \, dx \approx \frac{0.5}{2}\{(2.8182 + 8) + 2(4 + 5.1216 + 6.1716 + 7.1364)\}$$
$$= 13.91935 = 13.92 \text{ to 4 s.f.}$$
[4 marks available — 1 mark for the correct value of h, 1 mark for correct x- and y-values, 1 mark for using the formula correctly, 1 mark for the correct answer]

7 a) $n = 5$, $h = \dfrac{3 - 2.5}{5} = 0.1$

Work out the x- and y-values (y-values given to 6 s.f.):

$x_0 = 2.5$ $y_0 = 0.439820$
$x_1 = 2.6$ $y_1 = 0.424044$
$x_2 = 2.7$ $y_2 = 0.408746$
$x_3 = 2.8$ $y_3 = 0.393987$
$x_4 = 2.9$ $y_4 = 0.379802$
$x_5 = 3.0$ $y_5 = 0.366204$

$$\int_{2.5}^{3} y\,dx \approx \frac{0.1}{2}\{(0.439820 + 0.366204) + 2(0.424044$$
$$+\, 0.408746 + 0.393987 + 0.379802)\}$$
$$= 0.2009591... = 0.2010 \text{ (4 s.f.)}$$

[4 marks available — 1 mark for the correct value of h, 1 mark for correct x- and y-values, 1 mark for using the formula correctly, 1 mark for the correct answer]

b) The area of a rectangle with base 0.5 and height f(2.5) will be an overestimate of area R as the top of the rectangle will be above the top of the curve.
This rectangle has area $0.5 \times 0.439820 = 0.21991$ *[1 mark]*.
The area of a rectangle with base 0.5 and height f(3) will be an underestimate as the top of the rectangle will be below the top of the curve.
This rectangle has area $0.5 \times 0.366204 = 0.183102$ *[1 mark]*.
So the actual area of R lies between these two values, i.e. $0.183102 < R < 0.21991$. Both the upper and lower limits round to 0.2, so $R = 0.2$ correct to 1 d.p. *[1 mark]*

Pages 65-67: Vectors

1 a)

Use trigonometry to find the horizontal and vertical components of the magnitude $4\sqrt{2}$ vector:
Horizontal component: $4\sqrt{2}\cos 315° = 4\mathbf{i}$
Vertical component: $4\sqrt{2}\sin 315° = -4\mathbf{j}$
Resultant, $\mathbf{r} = 7\mathbf{j} + (4\mathbf{i} - 4\mathbf{j})$
So $\mathbf{r} = 4\mathbf{i} + 3\mathbf{j}$
[2 marks available — 1 mark for finding the horizontal and vertical components and 1 mark for correct expression for r]
Remember — angles are normally measured anticlockwise from the positive x-axis, so here, $360° - 45° = 315°$

b) $|\mathbf{r}| = \sqrt{4^2 + 3^2} = 5$
$\Rightarrow \mathbf{s} = 7\mathbf{r}$
So $\mathbf{s} = 7(4\mathbf{i} + 3\mathbf{j}) = 28\mathbf{i} + 21\mathbf{j}$
[3 marks available — 1 mark for finding magnitude of r, 1 mark for correct working to find s, 1 mark for correct expression for s]

2 $\overrightarrow{AB} = \overrightarrow{OB} - \overrightarrow{OA} = (5\mathbf{i} - 3\mathbf{j} + 6\mathbf{k}) - (-\mathbf{i} + 7\mathbf{j} - 2\mathbf{k})$
$= 6\mathbf{i} - 10\mathbf{j} + 8\mathbf{k}$ *[1 mark]*
$\overrightarrow{AM} = \frac{1}{2}\overrightarrow{AB} = 3\mathbf{i} - 5\mathbf{j} + 4\mathbf{k}$

Use this to find $\overrightarrow{CM}$:
$\overrightarrow{CM} = -\overrightarrow{OC} + \overrightarrow{OA} + \overrightarrow{AM}$
$= -(5\mathbf{i} + 4\mathbf{j} + 3\mathbf{k}) + (-\mathbf{i} + 7\mathbf{j} - 2\mathbf{k}) + (3\mathbf{i} - 5\mathbf{j} + 4\mathbf{k})$
$= -3\mathbf{i} - 2\mathbf{j} - \mathbf{k}$
[1 mark for a correct method, 1 mark for the correct vector]
$|\overrightarrow{CM}| = \sqrt{(-3)^2 + (-2)^2 + (-1)^2} = \sqrt{14}$
$|\overrightarrow{AB}| = \sqrt{6^2 + (-10)^2 + 8^2} = \sqrt{200}$
$k = \dfrac{|\overrightarrow{CM}|}{|\overrightarrow{AB}|} = \dfrac{\sqrt{14}}{\sqrt{200}}$ *[1 mark]*
$= \sqrt{\dfrac{7}{100}} = \dfrac{1}{10}\sqrt{7}$ *[1 mark]*

3 $\overrightarrow{AB} = \overrightarrow{OB} - \overrightarrow{OA}$
$= (14\mathbf{i} + 12\mathbf{j} - 9\mathbf{k}) - (-2\mathbf{i} + 4\mathbf{j} - 5\mathbf{k}) = (16\mathbf{i} + 8\mathbf{j} - 4\mathbf{k})$ *[1 mark]*
$\overrightarrow{AC} = \overrightarrow{OC} - \overrightarrow{OA}$
$= (2\mathbf{i} + \mu\mathbf{j} + \lambda\mathbf{k}) - (-2\mathbf{i} + 4\mathbf{j} - 5\mathbf{k}) = 4\mathbf{i} + (\mu - 4)\mathbf{j} + (\lambda + 5)\mathbf{k}$
[1 mark]
$\overrightarrow{AB}$ and $\overrightarrow{AC}$ both share the point A, so to show they're collinear you need to show that the vectors are parallel i.e. $\overrightarrow{AB} = k\,\overrightarrow{AC}$ for some constant k. So $(16\mathbf{i} + 8\mathbf{j} - 4\mathbf{k}) = k(4\mathbf{i} + (\mu - 4)\mathbf{j} + (\lambda + 5)\mathbf{k})$
[1 mark]
Equate coefficients of $\mathbf{i}$, $\mathbf{j}$, $\mathbf{k}$ separately:
$16 = 4k \Rightarrow k = 4$ *[1 mark]*
$8 = k(\mu - 4) \Rightarrow \mu = 6$
$-4 = k(\lambda + 5) \Rightarrow \lambda = -6$ *[1 mark for both μ and λ correct]*

4 $\overrightarrow{PR} = \overrightarrow{PQ} + \overrightarrow{QR} = \begin{pmatrix} 2 \\ -9 \end{pmatrix} + \begin{pmatrix} 14 \\ 6 \end{pmatrix} = \begin{pmatrix} 16 \\ -3 \end{pmatrix}$ *[1 mark]*
$|\overrightarrow{PQ}| = \sqrt{2^2 + (-9)^2} = \sqrt{85}$
$|\overrightarrow{QR}| = \sqrt{14^2 + 6^2} = \sqrt{232}$
$|\overrightarrow{PR}| = \sqrt{16^2 + (-3)^2} = \sqrt{265}$
[1 mark for attempting to find magnitudes using Pythagoras, 1 mark for all 3 magnitudes correct]
Find angle QPR using the cosine rule:
$\cos QPR = \dfrac{85 + 265 - 232}{2 \times \sqrt{85} \times \sqrt{265}} = 0.3931...$ *[1 mark]*
angle $QPR = \cos^{-1} 0.3931$
$= 66.851...° = 66.9°$ (3 s.f.) *[1 mark]*

5 $\overrightarrow{AB} = \overrightarrow{OB} - \overrightarrow{OA} = \begin{pmatrix} 4 \\ -12 \\ 8 \end{pmatrix} - \begin{pmatrix} 1 \\ -3 \\ 2 \end{pmatrix} = \begin{pmatrix} 3 \\ -9 \\ 6 \end{pmatrix}$ *[1 mark]*

D is $\frac{2}{3}$ of the way along $\overrightarrow{AB}$, so $\overrightarrow{AD} = \frac{2}{3}\overrightarrow{AB}$

$= \frac{2}{3}\begin{pmatrix} 3 \\ -9 \\ 6 \end{pmatrix} = \begin{pmatrix} 2 \\ -6 \\ 4 \end{pmatrix}$ *[1 mark]*

$\overrightarrow{OD} = \overrightarrow{OA} + \overrightarrow{AD} = \begin{pmatrix} 1 \\ -3 \\ 2 \end{pmatrix} + \begin{pmatrix} 2 \\ -6 \\ 4 \end{pmatrix} = \begin{pmatrix} 3 \\ -9 \\ 6 \end{pmatrix}$ *[1 mark]*

Now, $\overrightarrow{OD} = -\frac{1}{2}\overrightarrow{CE}$

$\Rightarrow \overrightarrow{CE} = -2\overrightarrow{OD} = -2\begin{pmatrix} 3 \\ -9 \\ 6 \end{pmatrix} = \begin{pmatrix} -6 \\ 18 \\ -12 \end{pmatrix}$ *[1 mark]*

So $\overrightarrow{OE} = \overrightarrow{OC} + \overrightarrow{CE} = \begin{pmatrix} -3 \\ 9 \\ -6 \end{pmatrix} + \begin{pmatrix} -6 \\ 18 \\ -12 \end{pmatrix} = \begin{pmatrix} -9 \\ 27 \\ -18 \end{pmatrix}$ *[1 mark]*

6 a) Position vector of drone A:
$\mathbf{a} + \mathbf{b} + \mathbf{c} = (4\mathbf{i} + 6\mathbf{j} + 5\mathbf{k}) + (-\mathbf{i} - 2\mathbf{j} - 2\mathbf{k}) + (-3\mathbf{j} + \mathbf{k})$
$= (3\mathbf{i} + \mathbf{j} + 4\mathbf{k})$ *[1 mark]*
Position vector of drone B:
$2\mathbf{a} + \mathbf{b} - 3\mathbf{c} = 2(4\mathbf{i} + 6\mathbf{j} + 5\mathbf{k}) + (-\mathbf{i} - 2\mathbf{j} - 2\mathbf{k}) - 3(-3\mathbf{j} + \mathbf{k})$
$= (7\mathbf{i} + 19\mathbf{j} + 5\mathbf{k})$ *[1 mark]*
So the distance between drones A and B is:
$\sqrt{(7-3)^2 + (19-1)^2 + (5-4)^2}$ *[1 mark]*
$= \sqrt{341} = 18.466... = 18.5$ m (3 s.f.) *[1 mark]*

b) Vector from drone A to drone B is:
$(7\mathbf{i} + 19\mathbf{j} + 5\mathbf{k}) - (3\mathbf{i} + \mathbf{j} + 4\mathbf{k}) = 4\mathbf{i} + 18\mathbf{j} + \mathbf{k}$
So the vector to take drone A to 2 m below drone B is:
$(4\mathbf{i} + 18\mathbf{j} + \mathbf{k}) - 2\mathbf{k} = 4\mathbf{i} + 18\mathbf{j} - \mathbf{k}$
[2 marks available — 1 mark for a correct method, 1 mark for the correct answer]

c) The position vector of drone B as it moves in the positive $\mathbf{j}$ direction can be described by:
$(7\mathbf{i} + (19 + \lambda)\mathbf{j} + 5\mathbf{k})$ m for $\lambda \geq 0$ *[1 mark]*
So at the limit of the drone's range:
$|7\mathbf{i} + (19 + \lambda)\mathbf{j} + 5\mathbf{k}| = 50$ m
$\Rightarrow \sqrt{7^2 + (19 + \lambda)^2 + 5^2} = 50$ *[1 mark]*
$\Rightarrow (19 + \lambda)^2 = 2500 - 49 - 25 = 2426$
$\Rightarrow \lambda^2 + 38\lambda + 361 = 2426$
$\Rightarrow \lambda^2 + 38\lambda - 2065 = 0$ *[1 mark]*

Solve for λ using the quadratic formula:

$$\lambda = \frac{-38 \pm \sqrt{38^2 - 4(1)(-2065)}}{2(1)} \quad \textit{[1 mark]}$$

$$\Rightarrow \lambda = \frac{-38 \pm \sqrt{9704}}{2}$$

$\Rightarrow \lambda = 30.25444...$ (ignore the $-$ve solution as $\lambda \geq 0$)

$\Rightarrow \lambda = 30.3$ (3 s.f) *[1 mark]*

OK I won't lie, that last part was pretty nasty. But actually... once you've got the initial equation written down, it's just standard algebra that you've been doing since GCSE.

Section Two — Statistics

Pages 68-73: Data Presentation and Interpretation

1 a) mean $= \dfrac{\sum x}{n} = \dfrac{500}{10} = 50$ *[1 mark]*

variance $= \dfrac{\sum x^2}{10} - 50^2 = \dfrac{25\,622}{10} - 2500$ *[1 mark]* $= 62.2$

So standard deviation $= \sqrt{62.2} = 7.89$ *[1 mark]*

b) The mean will be unchanged *[1 mark]*, because the new value is equal to the original mean *[1 mark]*.

c) The mean for the second group is higher than the mean for the first group, suggesting that scores for the second group of friends are generally higher. The standard deviation for the second group is lower, which means that there was less variation in the scores for this group than for the first group.
[2 marks available — 1 mark for a sensible comment comparing the means and 1 mark for a sensible comment comparing the standard deviations]

2 a) E.g. The proportion of people with an exercise rating of '1' decreased from 58.8% in 2001 to 56.5% in 2011, and the proportion of people with an exercise rating of '4' increased from 1.9% in 2001 to 2.9% in 2011.
[1 mark for a sensible comment using data from the table]

b) E.g. This may not be appropriate for this data set, because the data only shows how people are travelling for the longest part of their journey. It is possible that someone in the 'On foot' category actually walks less than someone in the 'Public motorised' category (e.g. if the walk to a train station or bus stop makes up a significant part of their journey to work).
[1 mark for a sensible comment]

3 a) The $0 \leq b < 10$ class has a width of 10 and a frequency of 12, so its frequency density is $12 \div 10 = 1.2$. Use this to label the vertical axis and draw the missing bars:

$10 \leq b < 15$ class: frequency density $= 23 \div 5 = 4.6$

$25 \leq b < 35$ class: frequency density $= 6 \div 10 = 0.6$

Then use the frequency density axis and the formula frequency = frequency density × class width to work out the missing values in the table:

Monthly phone bill, £b	Frequency
$0 \leq b < 10$	12
$10 \leq b < 15$	23
$15 \leq b < 18$	$5 \times 3 = 15$
$18 \leq b < 20$	$5 \times 2 = 10$
$20 \leq b < 25$	18
$25 \leq b < 35$	6

[4 marks available — 1 mark for using the frequency density and the $0 \leq b < 10$ bar or the $20 \leq b < 25$ bar to work out the scale on the vertical axis, 1 mark for both bars correct, 1 mark for using frequency = frequency density × class width, 1 mark for both entries in table correct]

b) £12.50 is halfway through the $10 \leq b < 15$ class, so there are approximately $23 \div 2 = 11.5$ students that pay between £12.50 and £15. £17.50 is $\frac{5}{6}$ of the way through the $15 \leq b < 18$ class, so there are $\frac{5}{6} \times 15 = 12.5$ students that pay between £15 and £17.50. So an estimate for the number of students that have a monthly phone bill of between £12.50 and £17.50 is $11.5 + 12.5 = 24$.
[2 marks available — 1 mark for a correct method, 1 mark for the correct answer]

c) Add columns to the table showing the class midpoint and the midpoint × frequency:

Monthly phone bill, £b	Frequency, f	Class midpoint, x	$f \times x$
$0 \leq b < 10$	12	5	60
$10 \leq b < 15$	23	12.5	287.5
$15 \leq b < 18$	15	16.5	247.5
$18 \leq b < 20$	10	19	190
$20 \leq b < 25$	18	22.5	405
$25 \leq b < 35$	6	30	180

There are $12 + 23 + 15 + 10 + 18 + 6 = 84$ students in total, and $\sum fx = 60 + 287.5 + 247.5 + 190 + 405 + 180 = 1370$.

So mean $= \dfrac{\sum fx}{\sum f} = \dfrac{1370}{84} = 16.309... = £16.31$ to 2 d.p.

[2 marks available — 1 mark for use of class midpoints, 1 mark for the correct answer]
You'll get full marks for this if you do your working on a calculator.

d) Add columns for x^2 and fx^2:

Monthly phone bill, £b	Frequency, f	x^2	$f \times x^2$
$0 \leq b < 10$	12	25	300
$10 \leq b < 15$	23	156.25	3593.75
$15 \leq b < 18$	15	272.25	4083.75
$18 \leq b < 20$	10	361	3610
$20 \leq b < 25$	18	506.25	9112.5
$25 \leq b < 35$	6	900	5400

$\sum fx^2 = 26\,100$

Standard deviation $= \sqrt{\dfrac{\sum fx^2}{\sum f} - \overline{x}^2} = \sqrt{\dfrac{26\,100}{84} - 16.309...^2}$

$= 6.68683... = £6.69$ (2 d.p.)

[2 marks available — 1 mark for $\sum fx^2$, 1 mark for the correct answer]
You'll get full marks for this if you do your working on a calculator.

e) The 20th percentile is in the $\dfrac{20}{100} \times 84 = 16.8$th position.

This lies in the class $10 \leq b < 15$.

So 20th percentile $= 10 + 5 \times \dfrac{16.8 - 12}{23} = 10 + 1.0434...$

$= £11.04$ (2 d.p.)

[3 marks available — 1 mark for calculating the position of the 20th percentile, 1 mark for substituting into the formula for linear interpolation, 1 mark for the correct answer]

4 a) Local Authority A = Camden
Local Authority B = Eastbourne
E.g. A has a higher percentage of people aged 0 to 44 than B and a lower percentage aged 45 and above, so you'd expect A to be the local authority with the lower mean age (i.e. Camden).
OR B has a higher percentage of people in both the youngest and oldest age groups while A has a higher percentage of data in the centre of the data, so you'd expect B to be the local authority with the higher standard deviation (i.e. Eastbourne).
[2 marks available — 1 mark for correctly identifying the Local Authorities (with explanation), 1 mark for sensible reasoning]

   b) Outliers are below $42.8 - (2 \times 23.7) = -4.6$ years (impossible) or above $42.8 + (2 \times 23.7) = 90.2$ years. *[1 mark]*
Since approximately 1.5% of data values are in the range 90+ for Eastbourne, it's likely that some of these are greater than 90.2, meaning the Eastbourne data is likely to have outliers at this upper end. *[1 mark]*

   c) E.g. There is a large number of data values, so a small number of outliers won't affect the value of the mean very much. *[1 mark]*

5 a) Q_2 position is $30 \div 2 = 15$, Q_1 position is $30 \div 4 = 7.5$, and Q_3 position is $(3 \times 30) \div 4 = 22.5$.
So, using the graph, $Q_2 \approx 3.8$ m *[1 mark]*, $Q_1 \approx 3.2$ m *[1 mark]*, and $Q_3 \approx 4.55$ m *[1 mark]*.
The interquartile range is $4.55 - 3.2 = 1.35$ m *[1 mark]*.

   b) E.g. The giraffes in the zoo are generally taller, as they have a higher median. Two giraffes in the game reserve have extreme heights that are outliers, but there are no outliers in the zoo. The two populations seem similarly varied, since they have similar ranges (the range for the zoo is 2.86 m and, ignoring the outliers, the range for the game reserve is 2.8 m), although the IQR for the giraffes in the zoo is slightly greater than for the giraffes in the game reserve.
[3 marks available — 1 mark for each of three correct comparisons]

6 a) E.g. The data is discrete. / The stem and leaf diagram shows the distribution of the data. / There aren't too many values to display. *[1 mark for a sensible explanation]*

   b) n is even and $\frac{n}{2} = 6$, so take the average of the 6th and 7th values. So the median (Q_2) is:
$\frac{1}{2}(5500 + 5800) = £5650$ *[1 mark]*.
Since $12 \div 4 = 3$, the lower quartile is the average of the 3rd and 4th values. So the lower quartile (Q_1) is:
$\frac{1}{2}(4200 + 4600) = £4400$ *[1 mark]*.
Since $12 \div 4 \times 3 = 9$, the upper quartile is the average of the 9th and 10th values. So the upper quartile (Q_3) is:
$\frac{1}{2}(6000 + 6200) = £6100$ *[1 mark]*.
The interquartile range is $6100 - 4400 = £1700$ *[1 mark]*.

   c) The upper limit for outliers can be defined as $Q_3 + 1.5 \times IQR$. For this data, $Q_3 + 1.5 \times IQR = 6100 + 1.5 \times 1700 = £8650$. £9100 is above the upper limit, so it is an outlier.
[1 mark for a sensible answer and calculation]
It's not worth using mean ± 2 standard deviations as the limits here as you've already worked out the quartiles and IQR in b).

   d) E.g. the manager should not include this outlier in his analysis as it might be caused by an outside factor, e.g. Christmas, so including it would not accurately reflect normal sales. /
The manager should include this outlier in his analysis as it accurately reflects actual sales for this period, even if one week was unusually high. *[1 mark for a sensible comment]*

   e) E.g. the mean and standard deviation are affected by outliers, whereas the median and IQR are not. So as this set of data includes an outlier, the median and IQR are more useful. / The mean and standard deviation take into account all data values, so are more useful as they reflect the actual data.
[2 marks available — 1 mark for stating which values are more useful, 1 mark for a sensible explanation]

7 a) 280 minutes *[1 mark]*

   b) If the runners were in the 21-25 age group in 2000, then they would be in the 31-35 age group in 2010. The graph shows that the 31-35 age group's mean time to finish the race was higher in 2010 than the 21-25 age group's times in 2000, which would mean the runners were slower, so the graph does not suggest these runners' times improved.
[2 marks available — 1 mark for sensible explanation and 1 mark for the correct conclusion]

   c) E.g. That the same runners participated in the 2010 marathon and the 2000 marathon *[1 mark for a sensible comment]*.

   d) E.g. If there were different numbers of runners in each age category, the value calculated by Serj would not take this into account *[1 mark for a sensible explanation]*. To calculate a more accurate value, he should multiply the mean for each age category by the number of runners in that category, add these values up then divide by the total number of runners *[1 mark for a suitable method]*.

Pages 74-77: Probability

1 a) A and B are mutually exclusive, so $P(A \cap B) = 0$.
This means the circles don't intersect.
B and C are independent, so
$P(B \cap C) = P(B) \times (P(C) = 0.4 \times 0.3 = 0.12$
$P(A \cap C) = 0.06$
Use this to fill in the Venn diagram:

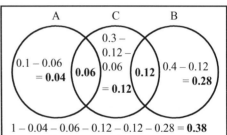

[5 marks available — 1 mark for circles overlapping correctly, 1 mark for calculating $P(B \cap C)$, 1 mark for calculating the remaining probabilities in both circles A and B, 1 mark for calculating the remaining probability in circle C, 1 mark for the correct probability outside the circles]

   b) $P(A) \times P(C) = 0.1 \times 0.3 = 0.03$
$P(A \cap C) = 0.06 \neq P(A) \times P(C)$,
so events A and C are not independent.
[2 marks available — 1 mark for finding $P(A) \times P(C)$ and stating $P(A \cap C)$, 1 mark for the correct conclusion]

   c) $P(B \cup C) = 0.06 + 0.12 + 0.12 + 0.28 = 0.58$ *[1 mark]*

   d) $P(A' \cap B') = 0.12 + 0.38 = 0.5$ *[1 mark]*

   e) $P(B'|A') = \dfrac{P(B' \cap A')}{P(A')} = \dfrac{0.5}{0.9} = \dfrac{5}{9} = 0.56$ (2 d.p.)
[2 marks available — 1 mark for using the conditional probability formula, 1 mark for the correct answer]
Use the numbers in the Venn diagram to help you answer parts c)-e).

2 As L and M are independent, $P(L \cap M) = P(L) \times P(M)$ *[1 mark]*
$P(L) = 0.1 + 0.4 = 0.5$ and $P(M) = 2x + 0.4$ *[1 mark]*
$P(L \cap M) = 0.4 = 0.5 \times (2x + 0.4) = x + 0.2 \Rightarrow x = 0.2$ *[1 mark]*
So this means $P(L \cup M) = 0.1 + 0.4 + 0.4 = 0.9$
$\Rightarrow P(L' \cap M') = 1 - 0.9 = 0.1$ *[1 mark]*

3   Draw a tree diagram:

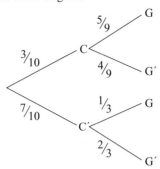

You need to find: $P(C|G) = \dfrac{P(C \cap G)}{P(G)}$ *[1 mark]*.

$P(G) = P(C \cap G) + P(C' \cap G)$:

$P(C \cap G) = \dfrac{3}{10} \times \dfrac{5}{9} = \dfrac{15}{90} = \dfrac{1}{6}$ *[1 mark]*,

and $P(C' \cap G) = \dfrac{7}{10} \times \dfrac{1}{3} = \dfrac{7}{30}$.

Adding these together you get: $P(G) = \dfrac{1}{6} + \dfrac{7}{30} = \dfrac{2}{5}$ *[1 mark]*.

So $P(C|G) = \dfrac{1}{6} \div \dfrac{2}{5} = \dfrac{5}{12}$ *[1 mark]*.

4   a)   Let U be the event 'selected fan is under 18'.
         There are 1945 males, of which 305 are under 18,
         so $P(U | M) = \dfrac{305}{1945} = \dfrac{61}{389}$.
         *[2 marks available — 1 mark for a correct method,
         1 mark for the correct answer]*
         *You could have found this using the formula for conditional
         probability.*

     b)   E.g. Alan has assumed that the fans at the next match will have
         the same mix of ages and genders, but many factors affect
         the attendance at a game, and different opponents may have a
         different age or gender balance in their fans.
         *[2 marks available — 1 mark for an assumption,
         1 mark for a sensible reason why this may not be valid]*

5   a)   Draw a tree diagram. Let C = 'pick a heart from the
         complete pack':

         The probability of at least one event happening is the same as
         1 minus the probability of neither event happening:
         $P(S \cup C) = 1 - P(S' \cap C')$
         $= 1 - (0.7 \times 0.75) = 1 - 0.525 = 0.475$
         *[2 marks available — 1 mark for a correct method,
         1 mark for the correct answer]*
         You could also get the answer by doing
         $P(S \cap C) + P(S \cap C') + P(S' \cap C)$.

     b)   Let P(H) be the probability of selecting at least two hearts, and
         let C_1, C_2 and C_3 be the event of selecting a heart with the first,
         second and third card respectively.
         Draw a tree diagram to calculate the probability of $P(H \cap C_1)$:

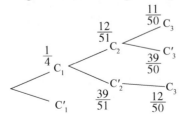

The branches of this tree diagram that are labelled with a
probability are the only ones where both a heart is selected first
and at least two hearts are selected, so:

$P(H \cap C_1) = \left(\dfrac{1}{4} \times \dfrac{12}{51} \times \dfrac{11}{50}\right) + \left(\dfrac{1}{4} \times \dfrac{12}{51} \times \dfrac{39}{50}\right)$
$+ \left(\dfrac{1}{4} \times \dfrac{39}{51} \times \dfrac{12}{50}\right) = \dfrac{89}{850}$

So, $P(H | C_1) = \dfrac{P(H \cap C_1)}{P(C_1)} = \dfrac{89}{850} \div \dfrac{1}{4} = \dfrac{356}{850}$
$= 0.419$ (3 s.f.)
*[3 marks available — 1 mark for a correct method to find
$P(H \cap C_1)$, 1 mark for $P(H \cap C_1)$ correct and 1 mark for the
correct answer]*
*You could also get this answer without using the formula, by just
considering the tree diagram from the point after C_1 has happened.*

     c)   E.g. I have assumed that Jessica is equally likely to select any
         of the cards in the complete pack of cards.
         [1 mark for any sensible comment]

6   $P(A | B) = 0.31 \Rightarrow \dfrac{P(A \cap B)}{0.4} = 0.31 \Rightarrow P(A \cap B) = 0.124$
    [1 mark]
    Now substitute this into the formula for $P(B | A)$:
    $P(B | A) = \dfrac{0.124}{P(A)} = 0.25 \Rightarrow P(A) = \dfrac{0.124}{0.25} = 0.496$ *[1 mark]*
    So, $P(A') = 1 - 0.496 = 0.504$ *[1 mark]*

7   a)   Draw a Venn diagram using the information you know:

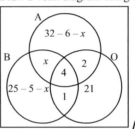

         [1 mark]
         Everything inside the circles should add up to 65:
         $(32 - 6 - x) + x + 4 + 2 + (25 - 5 - x) + 1 + 21 = 65$ *[1 mark]*
         $74 - x = 65 \Rightarrow x = 9$ *[1 mark]*
         So $x + 4 = 13$ children took an apple and a banana *[1 mark]*

     b)   Draw a tree diagram (X is the first child selects an apple,
         Y is the second child selects an apple):

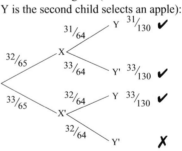

         The ticked outcomes are those where at least one person takes
         an apple. So $P(X \cup Y) = \dfrac{31}{130} + \dfrac{33}{130} + \dfrac{33}{130} = \dfrac{97}{130}$ *[1 mark]*
         $P(X | X \cup Y) = \dfrac{P(X \cap (X \cup Y))}{P(X \cup Y)}$ *[1 mark]*
         As all of X is contained within $X \cup Y$,
         $X \cap (X \cup Y)$ will be all of X. *[1 mark]*
         So $\dfrac{P(X \cap (X \cup Y))}{P(X \cup Y)} = \dfrac{P(X)}{P(X \cup Y)} = \dfrac{32}{65} \div \dfrac{97}{130} = \dfrac{64}{97}$ *[1 mark]*

Pages 78-84: Statistical Distributions

1   a)   The total probability must be 1, so go through all the possible
         values of x and add the probabilities:
         $\dfrac{k}{6} + \dfrac{2k}{6} + \dfrac{3k}{6} + \dfrac{3k}{6} + \dfrac{2k}{6} + \dfrac{k}{6} = \dfrac{12k}{6} = 2k$ *[1 mark]*.
         This must equal 1, so k must be $\dfrac{1}{2}$ *[1 mark]*.

     b) $P(1 < X \le 4) = P(X = 2) + P(X = 3) + P(X = 4)$ *[1 mark]*
         $= \dfrac{1}{6} + \dfrac{1}{4} + \dfrac{1}{4} = \dfrac{2}{3}$ *[1 mark]*

c) $P(3 \leq X < 6 \mid x \text{ odd}) = \dfrac{P(3 \leq X < 6 \text{ and } x \text{ odd})}{P(x \text{ odd})}$ *[1 mark]*

$P(3 \leq X < 6 \text{ and } x \text{ odd}) = P(X = 3) + P(X = 5)$
$= \dfrac{1}{4} + \dfrac{1}{6} = \dfrac{5}{12}$ *[1 mark]*

$P(x \text{ odd}) = P(X = 1) + P(X = 3) + P(X = 5)$
$= \dfrac{1}{12} + \dfrac{1}{4} + \dfrac{1}{6} = \dfrac{1}{2}$ *[1 mark]*

So $P(3 \leq X < 6 \mid x \text{ odd}) = \dfrac{5/12}{1/2} = \dfrac{5}{6}$ *[1 mark]*

2   Using $\displaystyle\sum_{\text{all } x} P(X = x) = 1$:

$0.4 + 0.3 + a + b = 1$ *[1 mark]* $\Rightarrow 0.7 + a + b = 1 \Rightarrow a + b = 0.3$
A contestant is twice as likely to be awarded 2 points as 3 points, so you also know that $a = 2b$. Substitute this in the above to get:
$2b + b = 0.3 \Rightarrow 3b = 0.3 \Rightarrow b = 0.1$ and $a = 2b = 0.2$
[1 mark for both correct]
P(one contestant scores 2 and the other scores 3)
$= P(\text{1st scores 2 and 2nd scores 3 or 1st scores 3 and 2nd scores 2})$
$= (0.2 \times 0.1) + (0.1 \times 0.2)$ *[1 mark]*
$= 0.04$ *[1 mark]*

3   a) The probability of getting 3 heads is:
$\dfrac{1}{2} \times \dfrac{1}{2} \times \dfrac{1}{2} = \dfrac{1}{8}$
The probability of getting 2 heads is: $3 \times \dfrac{1}{2} \times \dfrac{1}{2} \times \dfrac{1}{2} = \dfrac{3}{8}$
(multiply by 3 because any of the three coins could be the tail — the order in which the heads and the tail occur isn't important).
The probability of getting 1 or 0 heads
$= 1 - P(2 \text{ heads}) - P(3 \text{ heads}) = 1 - \dfrac{3}{8} - \dfrac{1}{8} = \dfrac{1}{2}$
Hence the probability distribution of X is:

x	20p	10p	nothing
$P(X = x)$	$\dfrac{1}{8}$	$\dfrac{3}{8}$	$\dfrac{1}{2}$

[3 marks for a completely correct table, otherwise 1 mark for drawing a table with the correct values of x and 1 mark for any one correct probability]

b) There are three ways to make a profit over two games:
win 20p and 10p, win 10p and 20p, or win 20p and 20p.
So P(makes profit over 2 games)
$= \left(\dfrac{1}{8} \times \dfrac{3}{8}\right) + \left(\dfrac{3}{8} \times \dfrac{1}{8}\right) + \left(\dfrac{1}{8} \times \dfrac{1}{8}\right)$
$= \dfrac{3}{64} + \dfrac{3}{64} + \dfrac{1}{64} = \dfrac{7}{64}$ *[1 mark]*
$P(\text{wins 40p}) = \dfrac{1}{8} \times \dfrac{1}{8} = \dfrac{1}{64}$ *[1 mark]*
P(wins 40p | makes profit over two games)
$= \dfrac{P(\text{wins 40p and makes profit})}{P(\text{makes profit})} = \dfrac{P(\text{wins 40p})}{P(\text{makes profit})}$ *[1 mark]*
$= \dfrac{1/64}{7/64} = \dfrac{1}{7}$ *[1 mark]*

4   a) 1) The probability P(chocolate bar contains a golden ticket) must be constant.
2) The trials must be independent (i.e. whether or not a chocolate bar contains a golden ticket must be independent of other chocolate bars).
[1 mark for each correct condition]

b) $P(X > 1) = 1 - P(X \leq 1)$ *[1 mark]*
$= 1 - 0.3990... = 0.601$ (3 s.f.) *[1 mark]*.

c) If more than 35 bars do not contain a golden ticket, then the student finds 4 or fewer golden tickets.
So find $P(X \leq 4)$ *[1 mark]* $= 0.952$ (3 s.f.) *[1 mark]*

5   a) Let the random variable X represent the number of cars in the sample of 20 that develop the rattle.
Then $X \sim B(20, 0.65)$ *[1 mark]*,
and you need to find $P(12 \leq X < 15)$.
$P(12 \leq X < 15) = P(X \leq 14) - P(X \leq 11)$
$= 0.7546... - 0.2376...$ *[1 mark for either]*
$= 0.517$ (3 s.f.) *[1 mark]*

b) $P(X > 10) = 1 - P(X \leq 10)$ *[1 mark]*
$= 1 - 0.1217...$
$= 0.8782... = 0.878$ (3 s.f.) *[1 mark]*.

c) The probability of more than half of a sample of 20 cars having the rattle is 0.8782... (using the full decimal from part b)).
Let Q be the number of samples containing more than 10 rattling cars. Then $Q \sim B(5, 0.8782...)$ *[1 mark]*.
$P(Q = 3) = \dbinom{5}{3} \times 0.8782...^3 \times (1 - 0.8782...)^2$ *[1 mark]*
$= 0.100$ (to 3 s.f.) *[1 mark]*
You could use your calculator's binomial functions here.

6   a) Let X be the number of customers in the sample of 20 who chose a sugar cone. Then $X \sim B(n, p)$ where $n = 20$, but you need to find p (the probability of choosing a sugar cone).
p is equal to the proportion of customers on the 1st July who chose a sugar cone, so:
$p = \dfrac{880}{1100} = 0.8$ *[1 mark]*, so $X \sim B(20, 0.8)$ *[1 mark]*
$P(X = 12) = \dbinom{20}{12} \times 0.8^{12} \times 0.2^8$ *[1 mark]*
$= 0.0222$ (3 s.f.) *[1 mark]*
You could use your calculator's binomial functions here.

b) Let Y be the number of customers who choose at least one scoop of chocolate ice cream.
Then $Y \sim B(75, 0.42)$ *[1 mark]*.
$P(Y > 30) = 1 - P(Y \leq 30)$ *[1 mark]*
$= 1 - 0.4099... = 0.590$ (3 s.f.) *[1 mark]*
You need to use your calculator's binomial c.d.f. to find P(Y ≤ 30).

c) E.g. it is not reasonable to assume that the customers' choices of ice cream are independent of each other, e.g. one person might see another person with a chocolate ice cream and decide to get one too, so the 75 trials are not independent. / The probability of choosing chocolate ice cream may not be constant, e.g. it might run out. Therefore the model may not be valid *[1 mark for a sensible comment]*.

7   a) E.g. The model is appropriate because (any two of):
- The histogram is roughly the shape of a normal distribution (i.e. bell-shaped).
- The histogram is roughly symmetrical about the mean.
- All the data lies between 2.0 kg and 5.0 kg (and 3 standard deviations from the mean is the range 2.05-5.05 kg).
[2 marks for stating the model is appropriate and giving two sensible reasons, otherwise 1 mark for a sensible comment]

b) Let M represent the masses of the newborn babies.
$M \sim N(3.55, 0.5^2)$
$P(M < 2.5) = 0.0179$ (3 s.f.)
[2 marks available — 1 mark for setting up the calculation correctly, 1 mark for the correct answer]

8   a) The mean lies halfway between the points of inflection,
so $\mu = \dfrac{5940 + 6030}{2} = 5985$ ml *[1 mark]*
The points of inflection occur at $\mu \pm \sigma$,
so $6030 = 5985 + \sigma \Rightarrow \sigma = 45$ ml *[1 mark]*
You could have found μ and σ by setting up and solving two simultaneous equations (μ + σ = 6030 and μ − σ = 5940), or by finding σ first.

b) $v \sim N(5985, 45^2)$
$P(5900 \leq v \leq 6100) = 0.965$ (3 s.f.)
[2 marks available — 1 mark for setting up the calculation correctly, 1 mark for the correct answer]

9   a) Let X represent the exam marks. Then $X \sim N(50, 15^2)$.
$P(X < 30) = 0.0912$ (3 s.f.)
[2 marks available — 1 mark for setting up the calculation correctly, 1 mark for the correct answer]

b) $P(X \geq 41) = 0.726$ (3 s.f.) *[1 mark]*
$0.726 \times 1000 = 726$, so approximately 726 candidates passed the exam *[1 mark]*

c) You need to find x such that $P(X \geq x) = 0.1$
Using the inverse normal function,
$x = 69$ marks (to the nearest whole number).
[2 marks available — 1 mark for a correct calculation, 1 mark for the correct answer]
If your calculator doesn't have an inverse normal function, you can convert to the standard normal distribution here and use the percentage-points table, with $p = 1 - 0.1 = 0.9$.

10 a) Let X represent the base diameters.
Then $X \sim N(12, \sigma^2)$. $P(X > 13) = 0.05$ *[1 mark]*,
so $P\left(Z > \dfrac{13 - 12}{\sigma}\right) = P\left(Z > \dfrac{1}{\sigma}\right) = 0.05$
$\Rightarrow P\left(Z \leq \dfrac{1}{\sigma}\right) = 1 - 0.05 = 0.95$
Using the percentage-points table, $p = 0.95$ for $z = 1.645$
$\Rightarrow \dfrac{1}{\sigma} = 1.645$ *[1 mark]*
$\Rightarrow \sigma = 1 \div 1.645 = 0.6079... = 0.608$ inches (3 s.f.) *[1 mark]*

b) $X \sim N(12, 0.6079...^2)$
$P(X < 10.8) = 0.02419...$ *[1 mark for setting up the calculation, 1 mark for the answer]*
So you would expect $0.02419... \times 100 \approx 2$ pizza bases to be discarded *[1 mark]*.

c) P(at least 1 base too small) = 1 − P(no bases too small)
P(base not too small) = $1 - 0.02419... = 0.9758...$
P(no bases too small) = $0.9758...^3 = 0.9291...$ *[1 mark]*
P(at least 1 base too small) = $1 - 0.9291...$ *[1 mark]*
$= 0.0708$ (3 s.f.) *[1 mark]*

11 a) $X \sim N(4, 1.1^2)$
$P(X < 3.5) = 0.325$ (3 s.f.)
[2 marks available — 1 mark for setting up the calculation correctly, 1 mark for the correct answer]

b) 'Deviates from the mean by more than 1 minute' means the window cleaner takes less than $4 - 1 = 3$ minutes or more than $4 + 1 = 5$ minutes, so you need to find:
$P(X < 3) + P(X > 5) = 0.1816... + 0.1816... = 0.363$ (3 s.f.)
[3 marks available — 1 mark for finding the values either side of the mean, 1 mark for setting up the calculation correctly, 1 mark for the correct answer]
Here, you could have worked out $1 - P(3 < X < 5)$ instead.

c) You need to find t such that $P(X > t) = 0.01$
Using the inverse normal function, $t = 6.6$ mins (1 d.p.)
(in minutes and seconds, this is 6 min and 36 seconds)
[2 marks available — 1 mark for a correct calculation, 1 mark for the correct answer]
If your calculator doesn't have an inverse normal function, you can convert to the standard normal distribution here and use the percentage-points table, with $p = 1 - 0.01 = 0.99$.

12 a) $M \sim N(93, \sigma^2)$ and $P(M \leq 95) = 0.75$ *[1 mark]*.
Transform this to a statement about the standard normal variable Z by subtracting the mean and dividing by the standard deviation:
$P\left(Z \leq \dfrac{95 - 93}{\sigma}\right) = 0.75 \Rightarrow P\left(Z \leq \dfrac{2}{\sigma}\right) = 0.75$
Using the percentage-points table: $\dfrac{2}{\sigma} = 0.674$ *[1 mark]*.
So $\sigma = \dfrac{2}{0.674} = 2.9673...$ *[1 mark]*.
So $M \sim N(93, 2.9673...^2)$, and you need to find $P(M < 88)$. *[1 mark]*
So the probability that a packet cannot be sold is
$P(M < 88) = 0.04599... = 0.0460$ (3 s.f.) *[1 mark]*.

b) First, you need to set up a binomial distribution for P, the number of packets that cannot be sold.
So $P \sim B(20, 0.04599...)$ *[1 mark]*. You need to find
$P(P = 1) = \dbinom{20}{1} \times 0.04599...^1 \times (1 - 0.04599...)^{19}$ *[1 mark]*
$= 0.376$ (3 s.f.) *[1 mark]*
Here, you had to spot that you needed to set up a binomial distribution — otherwise you'd have been a bit stuck.

13 a) S represents the number of sampled customers at Soutergate Cinema who went to see the superhero film, so $S \sim B(200, 0.48)$ *[1 mark]*.

b) A normal approximation is appropriate here because n is large (here, it's 200) *[1 mark]* and p is close to 0.5 (it's 0.48) *[1 mark]*.

c) The mean of a normal approximation to a binomial distribution is np and the variance is npq, where $q = 1 - p$:
$np = 200 \times 0.48 = 96$ and
$np(1 - p) = 200 \times 0.48 \times 0.52 = 49.92$
So $S \sim B(200, 0.48)$ can be approximated by $C \sim N(96, 49.92)$.
[2 marks available — 1 mark for a normal distribution with the correct mean, 1 mark for the correct variance]

d) A normal approximation would not be appropriate in this situation *[1 mark]*, as p is not close to 0.5 (it's 0.03) *[1 mark]*.
You could also say that $np = 90 \times 0.03 = 2.7 < 5$.

e) Let V represent the number of sampled customers at Vulcan Cinema who went to see the horror film. Then $V \sim B(90, 0.03)$.
$P(V > 4) = 1 - P(V \leq 4) = 1 - 0.8658... = 0.134$ (3 s.f.)
[3 marks available — 1 mark for setting up the correct binomial distribution, 1 mark for a correct calculation, 1 mark for the correct answer]

Pages 85-88: Statistical Hypothesis Testing

1 a) All the pupils in her school *[1 mark]*.

b) Opportunity (or convenience) sampling *[1 mark]*
It is unlikely to be representative because, e.g: the members of the sample are all in the same age group, so won't represent the whole school / older pupils are likely to have different views on politics to younger pupils / A-level politics students might be biased towards certain views / all the pupils who aren't in Josie's A-level politics class are excluded from the sample.
[1 mark for a sensible explanation]

c) E.g. Josie will only get responses from those pupils who choose to reply to her email / Those who make the effort to reply may be more likely to have strong views.
[1 mark for a sensible explanation]

2 a) E.g. people in different age groups may be likely to receive higher or lower pay rises, so it is important that the proportion in each age group in the sample is representative of the proportion in the population.
[1 mark for a sensible explanation]

b) Jamila can't use her data to draw conclusions about the whole population because, e.g: she only has data for one small town in one location / the types of employment in her town might not be representative of the whole of the UK / the age distribution of the working adults in Jamila's town might not be representative of the whole of the UK.
[1 mark for a sensible explanation]

c) E.g. She is more likely to sample people who work near the high street, who could receive higher or lower pay rises / the sample isn't random — she may choose to ask more younger people than older people, for example / the sample could exclude people who work on Monday afternoons.
[1 mark for a sensible explanation]

3 a) If p = proportion of authorities with fewer than 800 people who cycled to work, then $H_0: p = 0.2$ and $H_1: p \neq 0.2$ *[1 mark]*.
Let X = number of sampled authorities with fewer than 800 people who cycled to work.
Under H_0, $X \sim B(25, 0.2)$. *[1 mark]*
Under H_0, you'd expect there to be $25 \times 0.2 = 5$ authorities with fewer than 800 people who cycled to work. The observed value of 4 is less than this, so you're interested in the lower tail. Find the p-value of obtaining 4 or less:
$P(X \leq 4) = 0.421$ (3 s.f.) *[1 mark]*
It's a 2-tail test, so there's a probability of 0.05 in each tail.
$0.421 > 0.05$, so the result is not significant *[1 mark]*.
There is insufficient evidence at the 10% level to suggest that the proportion of authorities with fewer than 800 people who cycled to work was different for 2011 *[1 mark]*.

If you prefer, you can carry out the test by finding the critical region. This is calculated in part b).

b) The probability of incorrectly rejecting the null hypothesis equals the actual significance level of the test. The critical region for this test is in two parts.
The first part is $X \leq 1$, since $P(X \leq 1) = 0.02738...$ (and $P(X \leq 2) = 0.0982... > 0.05$) *[1 mark]*.
The second part is $X \geq 9$, since $P(X \geq 9) = 0.04677...$ (and $P(X \geq 8) = 0.109... > 0.05$) *[1 mark]*.
So the actual significance level and the probability of incorrectly rejecting the null hypothesis is $0.02738... + 0.04677... = 0.07416... = 0.0742$ (3 s.f.) *[1 mark]*.

4 a) If p = proportion of students who have done judo for at least two years, then H_0: $p = 0.2$ and H_1: $p \neq 0.2$. *[1 mark for both]*.
Let X = number of students in the sample of 20 who have done judo for at least two years. Under H_0, $X \sim B(20, 0.2)$, and you'd expect there to be $20 \times 0.2 = 4$ students in the sample who have done judo for at least two years. The observed value of 7 is more than this, so you're interested in the upper tail, i.e. when $X \geq 7$ *[1 mark for both $X \sim B(20, 0.2)$ and $X \geq 7$]*. Find the probability of a value of 7 or more:
$P(X \geq 7) = 1 - P(X \leq 6)$
$= 1 - 0.9133... = 0.0867$ (3 s.f.) *[1 mark]*
It's a 2-tail test, so there's a probability of 0.025 in each tail. $0.0867 > 0.025$, so the result is not significant *[1 mark]*.
There is insufficient evidence at the 5% level to suggest that the percentage who have done judo for at least two years is different *[1 mark]*.

You might have chosen to find the critical region instead of the p-value. Here's the working you'd need — you can generate the values one by one on your calculator, or you might be able to use your calculator to generate a table of values for the distribution.
It's a 2-tail test, so the critical region is split into two, with a probability of ≤ 0.025 in each tail.
For the lower tail:
$P(X \leq 0) = 0.0115$ (3 s.f.) and $P(X \leq 1) = 0.0692$ (3 s.f.)
For the upper tail:
$P(X \geq 9) = 1 - 0.9900... = 0.00998$ (3 s.f.) and $P(X \geq 8) = 1 - 0.9678... = 0.0321$ (3 s.f.)
So CR is $X = 0$ or $X \geq 9$.
7 does not lie in the critical region, so do not reject H_0.

b) E.g. One assumption is that the sample members are independent of each other, so that one member having done judo for at least two years doesn't affect whether another member has. This is likely to be true for Nate's test as he has chosen sample members randomly from all his students. A second assumption is that there is a constant probability of students having done judo for at least two years. This is likely to be true for Nate's test as by randomly selecting the sample from the whole population, he will avoid selecting only 'basic' or 'advanced' students.
[4 marks available — 1 mark each for two suitable assumptions, 1 mark each for two sensible comments that show that the assumptions are likely to be correct]

5 If p = proportion of gym members who use the pool, then H_0: $p = 0.45$ and H_1: $p < 0.45$ *[1 mark]*.
Let X be the number of people in a sample of 50 who use the pool. Then under H_0, $X \sim B(50, 0.45)$ and the result of the test is that the manager rejects H_0, so you're looking for the biggest possible value x such that $P(X \leq x) \leq 0.05$ *[1 mark]*.
Using a calculator, $P(X \leq 16) = 0.0427$ (3 s.f.) and $P(X \leq 17) = 0.0765$ (3 s.f.) *[1 mark for both probabilities]*.
So the maximum possible number in the sample who use the pool is 16 *[1 mark]*.

6 Let X represent the mass (in g) of a chocolate muffin and let μ = mean mass (in g) of the chocolate muffins.
Then H_0: $\mu = 110$ *[1 mark]* and H_1: $\mu < 110$ *[1 mark]*.
Under H_0, $X \sim N(110, 3^2)$, so $\overline{X} \sim N(110, 0.6)$.

Find the probability of a value of $\overline{X}$ at least as extreme as the observed value:
$P(\overline{X} < 108.5) = P\left(Z < \dfrac{108.5 - 110}{\sqrt{0.6}}\right)$
$= P(Z < -1.9364...) = 0.0264$ (3 s.f.)
[1 mark for both $\overline{X} \sim N(110, 0.6)$ and $\overline{X} < 108.5$, 1 mark for the correct probability]
$0.0264 < 0.1$ *[1 mark]*, so the result is significant *[1 mark]*.
There is sufficient evidence at the 10% level to suggest that the mean mass of the chocolate muffins is less than 110 g *[1 mark]*.
If you chose to find the critical region for this test, you should have found a CR of $Z < -1.28$ (3 s.f.).

7 Let X represent the height (in cm) of a sunflower in the second field and let μ = mean height (in cm) of the sunflowers in the second field.
Then H_0: $\mu = 150$ *[1 mark]* and H_1: $\mu \neq 150$ *[1 mark]*.
Under H_0, $X \sim N(150, 20)$, so $\overline{X} \sim N\left(150, \dfrac{20}{6}\right)$ *[1 mark]*.
$\overline{x} = 140$, so $z = \dfrac{140 - 150}{\sqrt{20}/\sqrt{6}} = -5.4772...$ *[1 mark]*
It's a 2-tail test at the 1% level, so the critical region is given by $P(Z > z) = 0.005$ or $P(Z < -z) = 0.005$. Using the percentage-points table, $P(Z \leq 2.575) = 0.995 \Rightarrow P(Z > 2.575) = 0.005$.
So the critical region is $Z < -2.575$ or $Z > 2.575$.
$-5.4772... < -2.575$ *[1 mark]*, so it lies in the critical region and the result is significant *[1 mark]*. There is sufficient evidence at the 1% level to suggest that the average height of the sunflowers in the second field is different from the first field *[1 mark]*.
If you chose to find the p-value instead of the critical region, you should have calculated $P(\overline{X} \leq 140)$ or $P(Z \leq -5.4772...) = 2.16023 \times 10^{-8}$, which is less than 0.005, so the result is significant.

8 a) Let μ = mean car wash duration (in minutes).
Then H_0: $\mu = 8$ *[1 mark]* and H_1: $\mu < 8$ *[1 mark]*.
Under H_0, $X \sim N(8, 1.2)$, so $\overline{X} \sim N(8, 0.06)$.
Find the probability of a value of $\overline{X}$ at least as extreme as the observed value:
$P(\overline{X} \leq 7.8) = P\left(Z \leq \dfrac{7.8 - 8}{\sqrt{0.06}}\right)$
$= P(Z \leq -0.8164...) = 0.207$ (3 s.f.)
[1 mark for both $\overline{X} \sim N(8, 0.06)$ and $\overline{X} \leq 7.8$, 1 mark for the correct probability]
$0.207 > 0.05$ *[1 mark]*, so the result is not significant *[1 mark]*.
There is insufficient evidence at the 5% level to suggest that the mean car wash duration has fallen *[1 mark]*.
The critical region for this test is $Z < -1.645$.

b) To fail to reject the null hypothesis requires z such that $P(Z < z) = 0.05$. Using the percentage-points table:
$P(Z \leq 1.645) = 0.95 \Rightarrow P(Z \leq -1.645) = 0.05$. So $z = -1.645$.
$z = \dfrac{\overline{x} - 8}{\sqrt{0.06}} = -1.645$ *[1 mark]* $\Rightarrow \overline{x} = 7.597...$
So the least value is 7.60 minutes (3 s.f.) *[1 mark]*.

9 The proportion of people who cycled to work in York is:
$\dfrac{10\,508}{87\,305} = 0.1203...$
If p is the proportion of people in the city outside of the UK who cycled to work, then H_0: $p = 0.1203...$, H_1: $p < 0.1203...$
Let X represent the number of people in the sample who cycled to work in the city outside the UK.
Then, under H_0, $X \sim B(1000, 0.1203...)$.
Approximate X by the normal distribution Y:
$\mu = np = 1000 \times 0.1203... = 120.3...$ and
$\sigma^2 = np(1-p) = 120.3... \times (1 - 0.1203...) = 105.8...$
So $Y \sim N(120.3..., 105.8...)$.
For there to be evidence to reject H_0 at the 2.5% significance level, $P(Y \leq c) < 0.025$. Since $P(Y \leq 100.19...) = 0.025$, $c \leq 100.19 \approx 100$, i.e. the largest possible value for c is approximately 100.

Or:

For the normal distribution approximately 95% of values lie in the range $\mu \pm 2\sigma$, so 2.5% of values are below $\mu - 2\sigma$ (the other 2.5% are above $\mu + 2\sigma$).
$\mu - 2\sigma = 120.3... - 2\sqrt{105.8...} = 99.7...$
So the largest value of $c \approx 100$.
[5 marks available — 1 mark for the correct value of p, 1 mark for stating the binomial distribution, 1 mark for attempting to find np and np(1 – p), 1 mark for a suitable method to find c, 1 mark for the correct value of c]
You could have used $\mu - 1.96\sigma = 100.1... \approx 100$ in the second method — which is slightly more accurate.

Pages 89-91: Correlation and Regression

1 a)

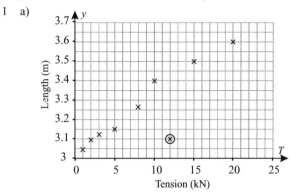

[1 mark for correctly circled point]

b) Positive (or strong positive) correlation — as the tension (T) increases, the length of the cable (y) also increases. *[1 mark]*

c) The missing data entries are:

T (kN)	1	3	10	15
y (metres)	3.05	3.13	3.4	3.5

[2 marks for all four values correct, otherwise 1 mark for at least two values correct]

d) a represents the length of the cable when it is not under tension, so when $T = 0$, the cable is 3 m long. *[1 mark]*
b represents the extension of the cable, so for every 1 kN increase in tension, the length of the cable increases by 0.03 m. *[1 mark]*

e) The estimate for 50 kilonewtons may be unreliable as it involves extrapolating beyond the range of the experimental data. *[1 mark]*

2 a) Positive (or weak positive) correlation — as the population size increases, the percentage of 16-74 year-olds not in employment also generally increases. *[1 mark]*

b)

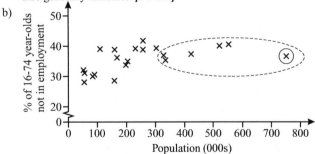

[1 mark for circling at least the point with a solid ring around it and no more than the points with a dotted ring around them]

3 a) E.g. the diagram does show negative correlation overall, so Jiao's claim seems reasonable *[1 mark]*. However, the data is clearly grouped into two clusters (possibly one for rural locations and one for urban locations), with little correlation within each cluster *[1 mark]*.

b) E.g. although correlation doesn't necessarily mean causation, in this case it seems reasonable that Killian is correct, as a habitat where reindeer thrive (e.g. a woodland) will not be densely populated — if it were, it would no longer be woodland so the deer population would decrease. / No, Killian is not likely to be correct — there could be a third factor affecting both populations (e.g. climate — reindeer prefer a colder climate, whereas population density for people is higher in more temperate regions.) *[1 mark for a suitable statement, 1 mark for a sensible explanation]*

4 Call the PMCC for the whole population ρ. Then H_0: $\rho = 0$, H_1: $\rho > 0$ *[1 mark for both]* (so this is a 1-tail test).
The test statistic is $r = 0.634$.
The significance level is $\alpha = 0.05$ and the sample size is 10.
Using the table provided, the critical value is 0.549 *[1 mark]*.
Since $0.634 > 0.549$ *[1 mark]*, the result is significant and there is evidence at the 5% level of significance to reject H_0 and to support the alternative hypothesis. There is evidence that the PMCC between floor area and advertised price of houses on the website is positive *[1 mark]*.

5 a) E.g. The outlier is similar to a lot of the data for Country A, so it's not an obvious measurement error, and so probably shouldn't be excluded from all the analysis. However, it may represent a very unusual city for Country B, so it might be sensible to exclude it from some analysis, depending on the analysis being done *[1 mark for a sensible explanation]*.

b) Call the PMCC for the whole population ρ. Then H_0: $\rho = 0$, H_1: $\rho \neq 0$ *[1 mark for both]* (so this is a 2-tail test).
The test statistic is $r = 0.833$ and the critical value is 0.765.
Since $0.833 > 0.765$ *[1 mark]*, the result is significant and there is evidence at the 1% level of significance to reject H_0 and to support the alternative hypothesis. There is evidence that the PMCC between population size and annual CO_2 emissions is non-zero for the 10 largest cities in country A *[1 mark]*.

Section Three — Mechanics

Pages 92-96: Kinematics 1

1 Displacement (S) is given by the area under the graph.
The area under the line is the area of a trapezium:
$$\text{area under the line} = \frac{U + V}{2} \times T$$
So $S = \frac{1}{2}(U + V)T$, as required.
[3 marks available — 1 mark for stating that displacement is the area under the graph, 1 mark for a using a suitable method to find the area under the line, 1 mark for the correct conclusion]

2 a) $u = 5$, $v = 0$, $a = -9.8$, $t = ?$
Using $v = u + at$: $0 = 5 - 9.8t \Rightarrow t = 5 \div 9.8 = 0.5102...$
$t = 0.510$ s (3 s.f.)
[3 marks available — 1 mark for using appropriate equation, 1 mark for correct workings, 1 mark for correct value of t]

b) $s = -2$, $u = 5$, $v = ?$, $a = -9.8$ Using $v^2 = u^2 + 2as$:
$v^2 = 5^2 + 2(-9.8 \times -2) = 64.2 \Rightarrow v = -8.0124...$
$= -8.01 \text{ ms}^{-1}$ (3 s.f.)
The velocity is negative when it reaches B as the ball is travelling in the negative direction (you can also see this from the graph).
[3 marks available — 1 mark for using appropriate equation, 1 mark for correct workings, 1 mark for correct value of v]

c) E.g. The graph shows the ball instantaneously changing velocity from point B to point C, which is unrealistic.
[1 mark for any sensible comment]

3 From passing the sign to entering the tunnel:
$s = ut + \frac{1}{2}at^2 \Rightarrow 110 = 8u + \frac{1}{2}(a \times 8^2)$ *[1 mark]*
$\Rightarrow 110 = 8u + 32a$ ① *[1 mark]*
From passing the sign to leaving the tunnel:
$s = ut + \frac{1}{2}at^2 \Rightarrow 870 = 32u + \frac{1}{2}(a \times 32^2)$ *[1 mark]*
$\Rightarrow 870 = 32u + 512a$ ② *[1 mark]*

② – 4 × ① gives: *[1 mark for a correct method]*
$430 = 384a \Rightarrow a = 1.119... = 1.12$ ms^{-2} (3 s.f.) *[1 mark]*
Substitute back into ① to find u:
$110 = 8u + 32a$
$u = \dfrac{110 - 32(1.119...)}{8} = 9.270... = 9.27$ ms^{-1} *[1 mark]*

4 a) $|\sqrt{1^2 + 3^2} - \sqrt{(-5)^2 + 2^2}| = |\sqrt{10} - \sqrt{29}| = 2.22$ ms^{-1} (3 s.f.)
[2 marks available — 1 mark for attempting to find the magnitude of each vector and 1 mark for the correct answer]

b) Draw a diagram:

So, angle is $180° - \tan^{-1}\left(\dfrac{2}{5}\right) = 158.2°$ (1 d.p.)
[2 marks available — 1 mark for attempting to find the direction of the vector and 1 mark for the correct answer]

c) Using $\mathbf{s} = \mathbf{v}t$, the position vector of A is:
$\mathbf{s} = (2\mathbf{i} + \mathbf{j}) + 8(\mathbf{i} + 3\mathbf{j}) = (2\mathbf{i} + \mathbf{j}) + (8\mathbf{i} + 24\mathbf{j})$
$= (10\mathbf{i} + 25\mathbf{j})$ m *[1 mark]*
Using $\mathbf{s} = \mathbf{v}t$, the position vector of B is:
$\mathbf{s} = (10\mathbf{i} + 25\mathbf{j}) + 5(-5\mathbf{i} + 2\mathbf{j}) = (10\mathbf{i} + 25\mathbf{j}) + (-25\mathbf{i} + 10\mathbf{j})$
$= (-15\mathbf{i} + 35\mathbf{j})$ m *[1 mark]*

5 a) Find the acceleration using the displacement at $t = 3$:
$\mathbf{u} = \mathbf{0}, \mathbf{a} = \mathbf{a}, \mathbf{s} = (9\mathbf{i} - 18\mathbf{j}), t = 3$
$\mathbf{s} = \mathbf{u}t + \dfrac{1}{2}\mathbf{a}t^2$
$(9\mathbf{i} - 18\mathbf{j}) = \dfrac{1}{2} \times \mathbf{a} \times 3^2$ *[1 mark]*
$(9\mathbf{i} - 18\mathbf{j}) = 4.5\mathbf{a} \Rightarrow \mathbf{a} = (2\mathbf{i} - 4\mathbf{j})$ ms^{-2} *[1 mark]*
The particle starts at O, so its position vector at time t is equal to its displacement:
So $\mathbf{p} = \mathbf{s} = \dfrac{1}{2}(2\mathbf{i} - 4\mathbf{j})t^2$ *[1 mark]*
$\mathbf{p} = (\mathbf{i} - 2\mathbf{j})t^2$ m *[1 mark]*

b) Find an expression for the second particle's displacement after t seconds:
$\mathbf{s} = \mathbf{v}t = (3\mathbf{i} - 5\mathbf{j})t$ *[1 mark]*
So the second particle's position vector at time t is:
$\mathbf{q} = (a\mathbf{i} + b\mathbf{j}) + (3\mathbf{i} - 5\mathbf{j})t$
When they collide, their position vectors are equal:
$(\mathbf{i} - 2\mathbf{j})t^2 = (a\mathbf{i} + b\mathbf{j}) + (3\mathbf{i} - 5\mathbf{j})t$
$t = 8$, so:
$64(\mathbf{i} - 2\mathbf{j}) = (a\mathbf{i} + b\mathbf{j}) + 8(3\mathbf{i} - 5\mathbf{j})$ *[1 mark]*
$64\mathbf{i} - 128\mathbf{j} - 24\mathbf{i} + 40\mathbf{j} = a\mathbf{i} + b\mathbf{j}$
$40\mathbf{i} - 88\mathbf{j} = a\mathbf{i} + b\mathbf{j}$
So $a = 40$ and $b = -88$ *[1 mark for both]*

6 a) The ball is travelling eastwards when the $\mathbf{j}$ component of the velocity vector is 0 and the $\mathbf{i}$ component is positive.
$3t - 8 = 0 \Rightarrow t = \dfrac{8}{3}$.
When $t = \dfrac{8}{3}$, the $\mathbf{i}$ component is $2\left(\dfrac{8}{3}\right) - 5 = \dfrac{1}{3} > 0$ *[1 mark]*,
so the ball is travelling eastwards at $t = \dfrac{8}{3}$ seconds *[1 mark]*.

b) Integrate $\mathbf{v}$ to find an expression for the ball's horizontal position (ignoring the $\mathbf{j}$ part of the calculation) :
$\int \mathbf{v_i}\, dt = \mathbf{x_i} = \int (2t - 5)\mathbf{i}\, dt$ *[1 mark]*
$= (t^2 - 5t)\mathbf{i} + \mathbf{c}$ *[1 mark]*
When $t = 2$, $\mathbf{x} = -2\mathbf{i} + 10\mathbf{j}$. Use this to find $\mathbf{c}$:
$(4 - 10)\mathbf{i} + \mathbf{c} = -2\mathbf{i} \Rightarrow \mathbf{c} = 4\mathbf{i}$ *[1 mark]*
So, $\mathbf{x_i} = (t^2 - 5t + 4)\mathbf{i}$.
The ball is west of the origin whenever
$t^2 - 5t + 4 = (t - 1)(t - 4) < 0$ *[1 mark]* $\Rightarrow 1 < t < 4$. *[1 mark]*

7 a) $s = \int v\, dt = \dfrac{11}{2}t^2 - \dfrac{2}{3}t^3 + c$ for $0 \leq t \leq 5$.
When $t = 0$, $s = 0 \Rightarrow c = 0$, so $s = \dfrac{11}{2}t^2 - \dfrac{2}{3}t^3$
So when $t = 5$:
$s = \dfrac{11}{2}(25) - \dfrac{2}{3}(125) = 54.16... = 54.2$ m (3 s.f.)
[4 marks available — 1 mark for integrating with respect to time, 1 mark for obtaining an expression for s, 1 mark for finding c, 1 mark for correct final answer]

b) $s = \int v\, dt$ *[1 mark]*
$= 25t - 2t^2 + k$ for $t > 5$ *[1 mark]*
When $t = 5$, $s = 54.16...$ (from part a)) — use this to find k:
$54.16... = 25(5) - 2(25) + k$. So, $k = -20.83...$ *[1 mark]*
Need to find t when $s = 0$, i.e. when:
$2t^2 - 25t + 20.83... = 0$ *[1 mark]*
Solve using quadratic formula: $t = 0.898$ or 11.6 s
$t = 0.898$ can be ignored, as it is outside the interval for which the equation is valid.
So P is back at the origin after 11.6 s (3 s.f.) *[1 mark]*

8 Integrate the acceleration to find an expression for the velocity.
a is a constant, so $v = \int a\, dt = at + c$ *[1 mark]*.
When $t = 0$, $v = u$, so $u = c$ and this gives $v = at + u$ *[1 mark]*.
Integrate again to find an expression for the object's displacement:
$s = \int v\, dt = \int at + u\, dt = \dfrac{1}{2}at^2 + ut + k$ *[1 mark]*.
The object sets off from the origin, so setting $t = 0$, $s = 0 \Rightarrow k = 0$
which gives $s = ut + \dfrac{1}{2}at^2$ as required *[1 mark]*.

9 a) $a = \dfrac{dv}{dt}$ *[1 mark]*
$a = 2 - 3(-2)e^{-2t} = (2 + 6e^{-2t})$ ms^{-2} *[1 mark]*

b) t must be greater than or equal to 0.
When $t = 0$, $a = 2 + 6e^{-2 \times 0} = 8$ ms^{-2} *[1 mark]*
As $t \to \infty$, $6e^{-2t} \to 0$, so $a \to 2$ ms^{-2} *[1 mark]*
So, $2 < a \leq 8$ *[1 mark]*

c) $s = \int v\, dt$ *[1 mark]*
$s = t^2 + \dfrac{3}{2}e^{-2t} + 4t + c$ *[1 mark]*
When $t = 0$, the particle is at the origin (i.e. $s = 0$):
$0 = 0 + \dfrac{3}{2} + c \Rightarrow c = -\dfrac{3}{2}$
So, $s = \left(t^2 + \dfrac{3}{2}e^{-2t} + 4t - \dfrac{3}{2}\right)$ m *[1 mark]*

10 a) $\mathbf{a} = \dot{\mathbf{v}} = [(-6\sin 3t + 5)\mathbf{i} + 2\mathbf{j}]$ ms^{-2}
[2 marks in total — 1 mark for attempting to differentiate the velocity vector, 1 mark for correctly differentiating both components]

b) The $\mathbf{j}$-component of the acceleration is constant, so the $\mathbf{i}$-component is the only one which can be maximised.
To do this, set $\sin 3t = -1$.
The minimum value of sin(anything) is −1 and because you're multiplying it by −6, this will give you the maximum value for the i-component.
$\mathbf{a} = ((-6 \times -1) + 5)\mathbf{i} + 2\mathbf{j} = 11\mathbf{i} + 2\mathbf{j}$ *[1 mark]*
Magnitude $= |\mathbf{a}| = \sqrt{11^2 + 2^2}$ *[1 mark]* $= \sqrt{121 + 4}$
$= \sqrt{125} = 5\sqrt{5}$ ms^{-2} *[1 mark]*

Pages 97-98: Kinematics 2

1 a) Consider vertical motion, taking up as positive:
$s = 0$; $u = p$; $a = -9.8$; $t = 5$;
Acceleration is constant, so can use a constant acceleration formula, e.g. use $s = ut + \dfrac{1}{2}at^2$
$0 = (5p) - 4.9(5^2)$ *[1 mark for using a suvat equation correctly]*
$5p = 122.5 \Rightarrow p = 24.5$ *[1 mark]*
You could also use v = u + at here, with v = 0; u = p; a = −9.8; and t = 2.5 (i.e. the time taken for the ball to reach its highest point, where vertical velocity is momentarily zero).

b) No acceleration horizontally, so use $s = ut$ *[1 mark]*
$s = 29 \times 5 = 145$ m *[1 mark]*

2 a) Need to find the times when the stone is 22 m above the ground (i.e. when it is at the level of projection), so consider vertical motion, taking up as +ve: *[1 mark]*
$s = 0$; $u = 14\sin 46°$; $a = -9.8$; $t = ?$
Acceleration is constant, so use a constant acceleration formula, e.g. $s = ut + \dfrac{1}{2}at^2$
$0 = 14\sin 46° t - 4.9t^2$ *[1 mark]*
$0 = (14\sin 46° - 4.9t)t$

So, $t = 0$ (i.e. when the stone is thrown) or
$14\sin46° - 4.9t = 0 \Rightarrow t = 2.055$ s *[1 mark]*
So the stone is at least 22 m above the ground for:
$2.055 - 0 = 2.06$ s (3 s.f.) *[1 mark]*

b) No acceleration horizontally, so the horizontal component of velocity remains constant at $14\cos 46°$ ms^{-1}. *[1 mark]*
Consider vertical motion once more, taking up as +ve to find the vertical component of the stone's final velocity:
$s = -22$; $u = 14\sin 46°$; $a = -9.8$; $v = ?$
$v^2 = u^2 + 2as = (14\sin 46°)^2 + (2 \times -9.8 \times -22)$ *[1 mark]*
$= 532.6...$ *[1 mark]*
Don't bother finding the square root, as you'd only have to square it again in the next bit of the answer.
Speed $= \sqrt{(14\cos 46)^2 + 532.6...}$ *[1 mark]*
$= 25.0$ ms^{-1} (3 s.f.) *[1 mark]*

3 a) Consider motion vertically, taking up as positive:
$s = ?$, $u = 10\sin 20°$, $v = 0$, $a = -9.8$. *[1 mark]*
Use $v^2 = u^2 + 2as$:
$0 = (10\sin 20°)^2 - 19.6s$ *[1 mark]*
$\Rightarrow s = (10\sin 20°)^2 \div 19.6 = 0.59682...$ m *[1 mark]*
The stone is thrown from 1 m above the ground, so the maximum height reached is $1 + 0.59682... = 1.60$ m (3 s.f.)
[1 mark]

b) Again consider vertical motion:
$s = -1$, $u = 10\sin 20°$, $a = -9.8$, $t = ?$.
Use $s = ut + \frac{1}{2}at^2$
$-1 = (10\sin 20°)t - 4.9t^2$ *[1 mark]*
$\Rightarrow 4.9t^2 - (10\sin 20°)t - 1 = 0$.
Use the quadratic formula to find t:
$t = \dfrac{10\sin 20° + \sqrt{(-10\sin 20°)^2 + (4 \times 4.9 \times 1)}}{9.8}$ *[1 mark]*
$= 0.91986... = 0.920$ s (3 s.f.) *[1 mark]*
You don't need to worry about the other value of t that the formula gives you as it will be negative, and the equation isn't valid for t < O.

c) E.g. Air resistance acting on the stone (affecting acceleration) has not been included in the model, so it may be inaccurate.
[1 mark for any suitable modelling assumption that may affect the accuracy of the model]

Pages 99-104 Forces and Newton's Laws

1 a) Resolve along x-axis:
$P\sin40° - 20\cos(105 - 90)° = 0$
so, $P = \dfrac{20\cos 15°}{\sin 40°} = 30.054... = 30.1$ N (3 s.f.)
[2 marks available in total — 1 mark for correct workings, 1 mark for correct value of P]

b) Resolve along y-axis:
$R = 30.054...\cos40° + 35 - 20\sin15°$
$= 52.8$ N (3 s.f.) up the y-axis
[3 marks available in total — 1 mark for correct workings, 1 mark for correct value of R and 1 mark for correct direction]
The horizontal component is zero, so the direction is along the y-axis.

2 a) If the resultant force is **R** then
$\mathbf{R} = \mathbf{A} + \mathbf{B} = (2\mathbf{i} - 11\mathbf{j}) + (7\mathbf{i} + 5\mathbf{j}) = (9\mathbf{i} - 6\mathbf{j})$ N *[1 mark]*
$|\mathbf{R}| = \sqrt{9^2 + (-6)^2}$ *[1 mark]* $= 10.8$ N (3 s.f.) *[1 mark]*

b) Using $\mathbf{F} = m\mathbf{a}$:
$(9\mathbf{i} - 6\mathbf{j}) = 0.5\mathbf{a} \Rightarrow \mathbf{a} = (18\mathbf{i} - 12\mathbf{j})$ ms^{-2}
[2 marks available — 1 mark for correct working, 1 mark for correct answer]

3 Resolving horizontally: $S\cos40° = F$ *[1 mark]*
Resolving vertically: $R = 2g + S\sin40°$ *[1 mark]*
The ring is stationary, so $F \leq \mu R$ *[1 mark]*.
So, $S\cos40° \leq 0.3(2g + S\sin40°)$ *[1 mark]*
$S\cos40° \leq 0.6g + 0.3S\sin40°$
$S\cos40° - 0.3S\sin40° \leq 0.6g$
$S(\cos40° - 0.3\sin40°) \leq 0.6g$
$S \leq 10.3$ N (3 s.f.) *[1 mark]*

4 a)

Resolving parallel to the plane ($\nearrow$):
$F_{net} = ma$
$25\cos15° - F - 7g\sin15° = 7 \times 0.2$ *[1 mark]*
$F = 25\cos15° - 7g\sin15° - 1.4$
$F = 4.993...$ N *[1 mark]*
Resolving perpendicular to the plane ($\nwarrow$):
$F_{net} = ma$
$R - 25\sin15° - 7g\cos15° = 7 \times 0$ *[1 mark]*
$R = 25\sin15° + 7g\cos15° = 72.732...$ N
Using $F = \mu R$: *[1 mark]*
$4.993... = \mu \times 72.732...$ *[1 mark]*
$\Rightarrow \mu = 0.0686... = 0.07$ (2 d.p.) *[1 mark]*

b)

Resolving perpendicular to the plane ($\nwarrow$):
$R = 7g\cos15°$ *[1 mark]*
$F = \mu R$ *[1 mark]*
$F = 0.0686... \times 7g\cos15°$
Resolving parallel to the plane ($\swarrow$):
$7g\sin15° - F = 7a$ *[1 mark]*
$7g\sin15° - (0.0686... \times 7g\cos15°) = 7a$
$a = 1.88...$ ms^{-2} *[1 mark]*
$s = 3$, $u = 0$, $a = 1.88...$, $t = ?$
$s = ut + \frac{1}{2}at^2$
$3 = \frac{1}{2} \times 1.88... \times t^2$ *[1 mark]*
$t^2 = 3.18... \Rightarrow t = 1.78$ s (3 s.f.) *[1 mark]*

5 a)

[1 mark]

b) Resolving horizontally:
$F = T\cos51.3°$ *[1 mark]*
so $T = \dfrac{1.5}{\cos 51.3°} = 2.40$ N (3 s.f.) *[1 mark]*
Resolving vertically:
$mg = R + T\sin51.3°$ *[1 mark]*
$R = \dfrac{F}{\mu} = 1.5 \div 0.6 = 2.5$ N *[1 mark]*
so $mg = 2.5 + 2.40\sin51.3°$
and $m = 0.4461... = 0.446$ kg (to 3 s.f.) *[1 mark]*

6

Resolving the forces parallel to the plane ($\nearrow$):
$W\sin\beta = F + T$ *[1 mark]*
As the system is in limiting equilibrium, $F = \mu R$, so
$W\sin\beta = \mu R + T$ *[1 mark]*
Resolving forces perpendicular to the plane ($\nwarrow$):
$W\cos\beta = R$ *[1 mark]*
Substitute this in for R in the above equation:
$W\sin\beta = \mu W\cos\beta + T$ *[1 mark]*
$\Rightarrow \mu W\cos\beta = W\sin\beta - T \Rightarrow \mu = \dfrac{W\sin\beta}{W\cos\beta} - \dfrac{T}{W\cos\beta}$
$= \tan\beta - \dfrac{T}{W}\sec\beta$
[1 mark for using trig identities to rearrange into required form]

7   Call the mass of the woman W.
Drawing two force diagrams will really help:

First, use $F = ma$ for the woman in the lift to find her mass:
$675 - Wg = 0.75W \Rightarrow 675 = 0.75W + Wg \Rightarrow W = 63.981...$ kg
Now use $F = ma$ for the whole connected system:
$T - (500 + 63.981...)g = 0.75(500 + 63.981...) \Rightarrow T = 5950$ N
[4 marks available — 1 mark for using $F = ma$ for the woman, 1 mark for finding the mass of the woman, 1 mark for using $F = ma$ on the whole system, 1 mark for finding the correct answer]

8   a)   The system is at rest, so equating the forces acting on A gives:
$35g - T = 0 \Rightarrow T = 35g$ *[1 mark]*
Now do the same for forces acting on B:
$Mg + K - T = 0 \Rightarrow K = T - Mg$
Substitute in the value of T to get:
$K = 35g - Mg = g(35 - M)$ *[1 mark]*

   b)   (i) $v = 1$, $u = 0$, $a = ?$ and $t = 3$. Using $v = u + at$:
$1 = 3a \Rightarrow a = \frac{1}{3}$ ms^{-2} *[1 mark]*
Using $F = ma$ for A: *[1 mark]*
$35g - T = 35 \times \frac{1}{3} \Rightarrow T = 35g - \frac{35}{3} = 331.33....$ *[1 mark]*
Using $F = ma$ for B:
$T - Mg = M \times \frac{1}{3} \Rightarrow T = M(g + \frac{1}{3})$ *[1 mark]*
$\Rightarrow M = 331.33... \div (g + \frac{1}{3}) = 32.7$ kg (3 s.f.) *[1 mark]*

   (ii)   E.g. The string is very long, B doesn't reach the pulley, A doesn't reach the ground, acceleration is constant.
[1 mark for a sensible assumption]

9   a)

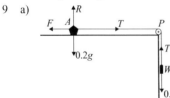

If mass of $A = 0.2$ kg, mass of $W = 1.5 \times 0.2 = 0.3$ kg
For W, $F = ma$:
$0.3g - T = 0.3(4)$ *[1 mark]*
$T = 2.94 - 1.2 = 1.74$ *[1 mark]*
For A, $F = ma$:
$T - F = 4(0.2) = 0.8$ *[1 mark]*
$F = \mu R$, but resolving vertically: $R = 0.2g = 1.96$ *[1 mark]*
So $F = 1.96\mu$, *[1 mark]*
so $T - 1.96\mu = 0.8$,
$\Rightarrow 1.74 - 1.96\mu = 0.8$
So $\mu = 0.480$ (3 s.f.) *[1 mark]*

   b)   Speed of A at h = speed of W at h (where it impacts ground).
Calculate speed of A at h using $v^2 = u^2 + 2as$:
$v^2 = 0^2 + 2(4 \times h)$ *[1 mark]*
$v^2 = 8h$ ①
Distance travelled by A beyond $h = \frac{3}{4}h$
Calculate frictional force slowing A after h using $F = ma$:
$F = \mu R$ and $R = mg = 0.2g = 1.96$ *[1 mark]*
so $F = 0.480 \times 1.96 = 0.94$
so $a = -0.94 \div 0.2 = -4.7$ ms^{-2} *[1 mark]*
Speed of A at h using $u^2 = v^2 - 2as$:
$u^2 = 3^2 - 2(-4.7 \times \frac{3}{4}h) = 9 + 7.05h$ *[1 mark]*   ②
Substituting ① into ② (where $v^2 = u^2$):
$8h = 9 + 7.05h$ so $h = 9.47$ m *[1 mark]*

Time taken to reach h using $s = ut + \frac{1}{2}at^2$:
$9.47 = (0 \times t) + \frac{1}{2}(4 \times t^2) = 2t^2$ *[1 mark]*
so $t^2 = 4.736...$ and $t = 2.18$ s (3 s.f.) *[1 mark]*
Tricky. The first thing is to realise that you can calculate the time taken for W to fall indirectly using the connected particle, A. The second thing to realise is that A carries on moving after W hits the ground and is slowed by friction. Then, find that you can calculate the speed of A when it's moved a distance of h in two different ways. Finally, you can find h and use that to find t.

   c)   E.g. There is no additional frictional force acting on the string to oppose motion. *[1 mark for a sensible comment]*

10 a)   Start by adding values to the diagram:

where $\theta = \tan^{-1}\frac{3}{4}$
$\tan\theta = \frac{3}{4} \Rightarrow \cos\theta = \frac{4}{5}$
Resolving forces around A perpendicular to the plane ($\nwarrow$):
$R = 3g\cos\theta = 3 \times 9.8 \times \frac{4}{5} = 23.52$ N *[1 mark]*
Use $F = ma$ for the vertical motion of B: *[1 mark]*
$(4 \times 9.8) - T = 4 \times 1.4$
So, $T = 33.6$ N *[1 mark]*
Use $F = ma$ for A parallel to the plane:
$T - F_r - (3 \times 9.8 \sin\theta) = 3 \times 1.4$ *[1 mark]*
$\tan\theta = \frac{3}{4} \Rightarrow \sin\theta = \frac{3}{5}$.
Substitute in values of T and $\sin\theta$:
$33.6 - F_r - 17.64 = 4.2$
So, $F_r = 11.76$ N *[1 mark]*
The system's moving, so friction is 'limiting',
i.e. $F_r = \mu R$ *[1 mark]*
Rearranging and substituting values in gives:
$\mu = \frac{F_r}{R} = \frac{11.76}{23.52} = 0.5$ *[1 mark]*

   b)   Motion of B: $u = 0$, $a = 1.4$, $s = ?$, $t = 2$
Use $s = ut + \frac{1}{2}at^2$: *[1 mark]*
$s = 0 + (\frac{1}{2} \times 1.4 \times 4) = 2.8$ m *[1 mark]*
Particle B moves 2.8 m before the string breaks.

   c)   While A and B are attached, they move together at the same speed. So you can use the information given to find the speed of (both) A and B when the string breaks:
$v = u + at = 0 + 1.4 \times 2 = 2.8$ ms^{-1} *[1 mark]*
Draw a diagram to show the forces on A after the string breaks:

Use $F = ma$ parallel to the plane to find a:
$-11.76 - (3 \times 9.8 \times \frac{3}{5}) = 3a$ *[1 mark]*
So, $a = -9.8$ ms^{-2} *[1 mark]*
Calculate the distance A travels using $v^2 = u^2 + 2as$ *[1 mark]*
$u = 2.8$, $v = 0$, $a = -9.8$, $s = ?$, $t = ?$
$s = \frac{v^2 - u^2}{2a} = \frac{0 - 2.8^2}{2 \times -9.8} = 0.4$ m *[1 mark]*
So particle A moves 0.4 m from the instant the string breaks until it comes to rest.

Pages 105-108: Moments

1 a) Start by drawing a diagram:

Take moments about A and B, using
moments clockwise = moments anticlockwise
A: $(20 \times 2) + (16 \times 3) = T_2 \times 4$
So, $T_2 = 88 \div 4 = 22$
B: $4 \times T_1 = (20 \times 2) + (16 \times 1)$
So, $T_1 = 56 \div 4 = 14$
[3 marks available — 1 mark for taking moments
about a point, 1 mark for correct value of A,
1 mark for correct value of B]
You might have resolved forces vertically or taken moments about
other points — if you get the same answers, and show full correct
working, then you're fine.

b) Start by drawing a new diagram:

Take moments about A and B:
A: $(20 \times 2) + 16d = 4 \times 2T_1$
So, $T_1 = 5 + 2d$
B: $4T_1 = (20 \times 2) + (4 - d) \times 16$
So, $T_1 = 10 + 16 - 4d = 26 - 4d$
Set the two equations in T_1 equal to each other and solve for d:
$5 + 2d = 26 - 4d \Rightarrow 6d = 21 \Rightarrow d = 3.5$ m
[3 marks available — 1 mark taking moments about a point,
1 mark for attempting to solve simultaneous equations,
1 mark for correct value of d]

2 a)

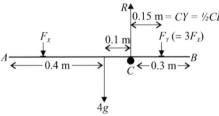

$CY = \frac{1}{2}CB = \frac{1}{2}(0.3) = 0.15$ m
$F_X + F_Y = 80g$, $F_Y = 3 \times F_X$,
so, $4F_X = 80g$, $F_X = 20g$ and $F_Y = 60g$
Taking moments about C:
$F_Y \times CY = (4g \times 0.1) + (F_X \times CX)$
so $(60g \times 0.15) = (4g \times 0.1) + (20g \times CX)$
and $CX = \dfrac{9g - 0.4g}{20g} = 0.43$ m

$AX = 0.5 - CX = 0.07$ m
[5 marks available — 1 mark for finding distances
CB and CY, 1 mark for finding correct values of F_X and F_Y,
1 mark for taking moments about a point, 1 mark for correct
workings, 1 mark for correct value of AX]

b) $F_X + F_Y = 80g$
Resolving vertically:
$R = F_X + F_Y + 4g = 84g = 823$ N (3 s.f.)
[2 marks available — 1 mark for resolving vertically, 1 mark
for correct value of R]

c)

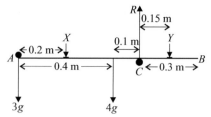

$AX = 0.07 + 0.13 = 0.2$ m
$F_X + F_Y = 80g$
Resolving vertically:
$R = 3g + F_X + 4g + F_Y = 87g$
Taking moments about point X:
$(0.2 \times 3g) + (0.3 \times 87g) = (0.2 \times 4g) + (0.45 \times F_Y)$
so $F_Y = 564$ N
and $F_X = 80g - 564 = 784 - 564 = 220$ N
OR:
Taking moments about point Y:
$(0.65 \times 3g) + (0.45 \times F_X) + (0.25 \times 4g) = (0.15 \times 87g)$
so $F_X = 220$ N
and $F_Y = 80g - 220 = 784 - 220 = 564$ N
[5 marks available — 1 mark for calculating new value
of AX, 1 mark for equation connecting F_X and F_Y, 1 mark for
taking moments around a point, 1 mark for correct value
of F_X, 1 mark for correct value of F_Y]

3 a)

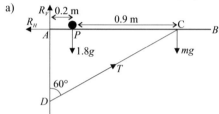

Taking moments about C:
$1.1 \times R_V = 0.9 \times 1.8g$
so, $1.1R_V = 15.876$
and $R_V = 14.43... = 14.4$ N (3 s.f.)
[3 marks available — 1 mark for taking moments about C,
1 mark for correct workings, 1 mark for correct value of R_V]
It's easy to get confused by which directions R_H and R_V act in.
My suggestion — make an educated guess and if you're wrong
they'll just turn out to be negative. No need to panic then.

b)

Magnitude: $|R| = \sqrt{14.43^2 + 35^2} = 37.9$ N (3 s.f.)
Angle to the horizontal: $\tan \theta = \dfrac{14.43}{35}$, so $\theta = 22.4°$ (3 s.f.)
[3 marks available — 1 mark for correct workings, 1 mark for
correct magnitude, 1 mark for correct angle]

c) Resolving horizontally:
$T\sin 60° = R_H = 35$
so, $T = \dfrac{35}{\sin 60°} = 40.4$ N (to 3 s.f.)

[3 marks available — 1 mark for resolving horizontally,
1 mark for correct workings, 1 mark for correct value of T]

4

Resolve all forces into components parallel and perpendicular to
the base of the shape, as shown above. Taking moments about A:
$4\sin 30° \times 2.28 + R\cos 30° \times 0 + R\sin 30° \times 4 = 4\cos 30° \times 2$
$2 \times 2.28 + 2R = 8\cos 30°$
$\Rightarrow R = (8\cos 30° \div 2) - 2.28 = 1.18$ N (3 s.f.)

[4 marks available in total — 1 mark for resolving, 1 mark for attempting to take moments, 1 mark for using correct values when taking moments, 1 mark for correct answer]

Don't be thrown too much by this being a lamina rather than a rod. Just approach it like you would any other moments question: resolve the forces into components, and then take moments using the perpendicular distance to the line of action of the force.

5 a)

Taking moments about A:
$54(3 - l) = 1.5 \times 8g\cos\theta$
$162 - 54l = 107.016$
so, $54l = 54.984 \Rightarrow l = 1.02$ m (to 3 s.f.)
[3 marks available — 1 mark for taking moments about A, 1 mark for correct workings, 1 mark for correct value of l]

b) Resolving vertically:
$R + 54\cos\theta = 8g$
so, $R = 8g - (54 \times 0.91) = 29.26$ N
Resolving horizontally: $F = 54\sin\theta$
Rod is in limiting equilibrium so $F = \mu R$:
$54\sin\theta = 29.26\mu$
As $\cos\theta = 0.91$, $\theta = 24.49...°$
$29.26\mu = 54\sin(24.49...°) = 22.38...$
So, $\mu = 0.765$ (3 s.f.)
[6 marks available — 1 mark for resolving vertically, 1 mark for correct value of R, 1 mark for resolving horizontally, 1 mark for using F = μR, 1 mark for correct workings, 1 mark for correct value of μ]

c)

Taking moments about C (and using the rod is about to rotate, so the reaction force from the ground is zero):
$$Vl\cos\theta = 8g\left(3 - l - \frac{3}{2}\right) = 8g\left(\frac{3}{2} - l\right)\cos\theta$$
$$\text{So } V = \frac{8g}{l}\left(\frac{3}{2} - l\right) = 4g\left(\frac{3}{l} - 2\right)$$

[2 marks available — 1 mark for equating the clockwise and anticlockwise moments about C, and 1 mark for rearranging correctly to give the required result]

6

Draw a diagram showing all the forces, including the friction and reaction forces where the ladder meets the ground (F_G and R_G), and the friction and reaction forces where the ladder meets the wall (F_W and R_W).
Resolving the forces vertically: $10g + 62g = R_G + F_W$
i.e. $R_G + F_W = 72g$
Resolving the forces horizontally: $F_G = R_W$
Since the ladder is in limiting equilibrium, you know that the frictional forces are given by:
$F_W = \mu_W R_W$ (where μ_W is the coefficient of friction between the ladder and the wall = 0.5), giving $F_W = 0.5R_W$, or $R_W = 2F_W$.

$F_G = \mu_G R_G$ (where μ_G is the coefficient of friction between the ladder and the ground = 0.4), giving $F_G = 0.4R_G$, or $R_G = 2.5F_G$.
Using the above equations:
$R_G + F_W = 72g$
$\Rightarrow 2.5F_G + F_W = 72g$
$\Rightarrow 2.5R_W + F_W = 72g$
$\Rightarrow 2.5 \times 2F_W + F_W = 72g$
or $6F_W = 72g$, which means $F_W = 12g$.
You can then work out that $R_W = 24g$, $F_G = 24g$, and $R_G = 60g$.
You could have worked out the reaction forces first, but the important thing is that you know how they relate to each other.
Now take moments about the bottom of the ladder:
Clockwise: $(10g\cos\theta \times 4) + (62g\cos\theta \times 6) = 412g\cos\theta$.
Anticlockwise: $F_W \times 8\cos\theta + R_W \times 8\sin\theta$
$= 12g \times 8\cos\theta + 24g \times 8\sin\theta$
$= 96g\cos\theta + 192g\sin\theta$.
Equating clockwise and anticlockwise moments:
$412g\cos\theta = 96g\cos\theta + 192g\sin\theta$.
This gives $316g\cos\theta = 192g\sin\theta$, which simplifies to
$\tan\theta = \dfrac{316}{192}$, or $\tan\theta = \dfrac{79}{48}$ (using $\tan\theta = \dfrac{\sin\theta}{\cos\theta}$).
[7 marks available — 1 mark for stating all the frictional and reaction forces to be considered, 1 mark for resolving the forces horizontally and vertically, 1 mark for giving the correct equations for the frictional or reaction forces, 1 mark for using the correct method to solve the simultaneous equations generated, 1 mark for finding all the necessary values from the simultaneous equations, 1 mark for taking moments about one end of the ladder, 1 mark for correctly rearranging to show the required result]

7 a)

Taking moments about A:
$(30\cos 70° \times 4) + 805\cos 70° \times (8 - x) = 260\sin 70° \times 8$
So $120 + 805(8 - x) = 2080\tan 70°$
(dividing through by $\cos 70°$).
This means $8 - x = (2080\tan 70° - 120) \div 805$
$8 - x = 6.95000...,$
or $x = 8 - 6.95000... = 1.04999...$ m.
This means $x = 1.05$ m (to the nearest cm).
[3 marks available — 1 mark for taking moments around A, 1 mark for correct workings, 1 mark for the correct value of x]

b) The horizontal force exerted on the bottom of the ladder, P, is directed towards the wall. The frictional force F between the ladder and the ground is also directed towards the wall. The reaction force R from the ground is directed vertically upwards.
Resolving vertically:
$805 + 30 = R$, so $R = 835$ N.
Resolving horizontally:
$P + F = 260$, where the maximum value of F (the limiting friction) is given by $F = \mu R = 0.1 \times 835 = 83.5$ N.
This means the minimum value of P needed is
$P = 260 - 83.5 = 176.5$ N.
[5 marks available — 1 mark for resolving the forces vertically, 1 mark for the correct value of R, 1 mark for resolving the forces horizontally, 1 mark for using the maximum value of F = μR, 1 mark for the correct minimum value of P]

c) If the ladder is uniform, you can assume all the weight acts at the centre. *[1 mark for correct assumption]*

158

Practice Exam Paper 1: Pure Mathematics

1 a) If $(x + 1)$ is a factor of f(x), then f(–1) = 0:
$f(-1) = (-1)^3 - 2(-1)^2 - 13(-1) - 10 = -1 - 2 + 13 - 10 = 0$
[1 mark], so $(x + 1)$ is factor of f(x) by the factor theorem
[1 mark].

b) $(x + 1)(x^2 - 3x - 10) = 0$ *[1 mark]*
$(x + 1)(x + 2)(x - 5) = 0$ *[1 mark]*
$\Rightarrow x = -2, -1$ and 5 *[1 mark]*

2 $2 \log a^3b - \log ab = \log (a^3b)^2 - \log ab$ *[1 mark]*
$= \log a^6b^2 - \log ab$
$= \log \dfrac{a^6 b^2}{ab}$ *[1 mark]*
$= \log a^5b$ *[1 mark]*

3 a) $y = \dfrac{2x + 7}{3x - 5} \Rightarrow y(3x - 5) = 2x + 7 \Rightarrow 3xy - 5y = 2x + 7$
$\Rightarrow 3xy - 2x = 7 + 5y \Rightarrow x(3y - 2) = 7 + 5y$
$\Rightarrow x = \dfrac{7 + 5y}{3y - 2}$
So $f^{-1}(x) = \dfrac{7 + 5x}{3x - 2}$
[2 marks available — 1 mark for a correct method, 1 mark for the correct answer]

b) $f(2) = \dfrac{2(2) + 7}{3(2) - 5} = \dfrac{4 + 7}{6 - 5} = \dfrac{11}{1} = 11$
$g(11) = 11^2 - k$, so $120 = 121 - k \Rightarrow k = 1$
[3 marks available — 1 mark for correct method for finding f(2), 1 mark for correct value of f(2), 1 mark for the correct value of k]

c) $gh(x) = (\cos x)^2 - 1 = \cos^2 x - 1$ *[1 mark]*
The range of $\cos^2 x$ is $0 \leq \cos^2 x \leq 1$,
so the range of gh(x) is $-1 \leq gh(x) \leq 0$ *[1 mark]*
$hg(x) = \cos(x^2 - 1)$ *[1 mark]*
So the range of hg(x) is the range of cos,
i.e. $-1 \leq hg(x) \leq 1$ *[1 mark]*

4 $\tan 3\theta \approx 3\theta$, $\cos 4\theta \approx 1 - \dfrac{16\theta^2}{2} = 1 - 8\theta^2$ and $\sin \theta^2 \approx \theta^2$, so:

$\dfrac{1 - \tan 3\theta}{\cos 4\theta - \sin \theta^2} \approx \dfrac{1 - 3\theta}{1 - 8\theta^2 - \theta^2} = \dfrac{1 - 3\theta}{1 - 9\theta^2} = \dfrac{1 - 3\theta}{(1 + 3\theta)(1 - 3\theta)}$

$= \dfrac{1}{1 + 3\theta}$ as required

[3 marks available — 1 mark for substituting the correct small angle approximations into the given function, 1 mark for factorising the denominator, 1 mark for correct answer]

5 a) $5u_2 - 2 = 38 \Rightarrow 5u_2 = 40 \Rightarrow u_2 = 8$, so
$5u_1 - 2 = 8 \Rightarrow 5u_1 = 10 \Rightarrow u_1 = 2$.
[2 marks available — 1 mark for correct working, 1 mark for correct answer]

b) For a decreasing sequence, $5u_n - 2 < u_n$ for all integer n
$\Rightarrow 4u_n < 2$ for all integer $n \Rightarrow u_n < \dfrac{1}{2}$ for all integer n.
This has to hold for $n = 1$, so $u_1 < \dfrac{1}{2}$.
[2 marks available — 1 mark for correct working, 1 mark for correct answer]

c) $\sum\limits_{n=1}^{r} u_n = \dfrac{1}{2}r$, means that every value of u_n must be $\dfrac{1}{2}$, including u_1, so $u_1 = \dfrac{1}{2}$ *[1 mark]*.

6 Proof by contradiction: assume that $\sqrt{2}$ is rational and can be written $\dfrac{a}{b}$, where a and b are both non-zero integers and have no common factors *[1 mark]*.
$\sqrt{2} = \dfrac{a}{b} \Rightarrow \sqrt{2}b = a \Rightarrow 2b^2 = a^2$ *[1 mark]*
a^2 must be even, so a must also be even, which means it can be written as $2k$, where k is an integer:
$2b^2 = (2k)^2 = 4k^2 \Rightarrow b^2 = 2k^2$ *[1 mark]*
This means that b^2 is even, which means that b is also even so can be written as $2l$, where l is an integer. This shows that a and b have a common factor of 2, which contradicts the initial assumption, so $\sqrt{2}$ must be an irrational number *[1 mark]*.

7 a) $\dfrac{11x - 7}{(2x - 4)(x + 1)} \equiv \dfrac{A}{2x - 4} + \dfrac{B}{x + 1}$ *[1 mark]*
$\Rightarrow 11x - 7 \equiv A(x + 1) + B(2x - 4)$ *[1 mark]*
Putting in $x = -1$ gives: $-18 = -6B \Rightarrow B = 3$
Putting in $x = 2$ gives: $15 = 3A \Rightarrow A = 5$
So $\dfrac{11x - 7}{(2x - 4)(x + 1)} \equiv \dfrac{5}{2x - 4} + \dfrac{3}{x + 1}$.
[1 mark for each correct fraction]

b) $\dfrac{11x - 7}{(2x - 4)(x + 1)} \equiv 5(2x - 4)^{-1} + 3(x + 1)^{-1}$ *[1 mark]*

$5(2x - 4)^{-1} = 5\left((-4)^{-1}\left(-\dfrac{x}{2} + 1\right)^{-1}\right)$

$= 5\left(-\dfrac{1}{4}\left(1 + -1\left(-\dfrac{x}{2}\right) + \dfrac{-1 \times -2}{1 \times 2}\left(-\dfrac{x}{2}\right)^2 + ...\right)\right)$ *[1 mark]*

$= -\dfrac{5}{4} - \dfrac{5}{8}x - \dfrac{5}{16}x^2 + ...$

$3(x + 1)^{-1} = 3\left(1 + -1(x) + \dfrac{-1 \times -2}{1 \times 2}x^2 + ...\right)$ *[1 mark]*

$= 3 - 3x + 3x^2 + ...$

$\dfrac{11x - 7}{(2x - 4)(x + 1)} \equiv -\dfrac{5}{4} - \dfrac{5}{8}x - \dfrac{5}{16}x^2 + 3 - 3x + 3x^2 + ...$

$\equiv \dfrac{7}{4} - \dfrac{29}{8}x + \dfrac{43}{16}x^2 + ...$

[1 mark for each correct term]

8 a) Rearrange the parametric equations:
$x - 1 = 5 \cos t \Rightarrow (x - 1)^2 = 25 \cos^2 t$ *[1 mark]*
$y - 2 = 5 \sin t \Rightarrow (y - 2)^2 = 25 \sin^2 t$ *[1 mark]*
$(x - 1)^2 + (y - 2)^2 = 25 \cos^2 t + 25 \sin^2 t$
$= 25(\sin^2 t + \cos^2 t) = 25$ *[1 mark]*

b) Gradient of the radius from the centre (1, 2) to (4, 6):
$\dfrac{6 - 2}{4 - 1} = \dfrac{4}{3}$ *[1 mark]*
So the gradient of the tangent at P is $-1 \div \dfrac{4}{3} = -\dfrac{3}{4}$ *[1 mark]*
Using the equation of a straight line:
$y - 6 = -\dfrac{3}{4}(x - 4)$ *[1 mark]* $\Rightarrow 3x + 4y - 36 = 0$ *[1 mark]*
(so $a = 3$, $b = 4$ and $c = -36$)

9 a) Use the product rule: $u = x \Rightarrow \dfrac{du}{dx} = 1$
$v = \ln x \Rightarrow \dfrac{dv}{dx} = \dfrac{1}{x}$ *[1 mark for both correct]*
$\dfrac{dy}{dx} = x\left(\dfrac{1}{x}\right) + \ln x = 1 + \ln x$ *[1 mark]*
At the stationary point, $\dfrac{dy}{dx} = 0$
$1 + \ln x = 0$ *[1 mark]* $\Rightarrow \ln x = -1 \Rightarrow x = e^{-1}$ *[1 mark]*
$y = e^{-1} \times \ln e^{-1} = -e^{-1}$
So the stationary point is $(e^{-1}, -e^{-1})$ *[1 mark]*

b) $\dfrac{d^2y}{dx^2} = \dfrac{1}{x}$ *[1 mark]*
At $(e^{-1}, -e^{-1})$, $\dfrac{d^2y}{dx^2} = e > 0$, so it is a minimum point *[1 mark]*

10 a) Let M be the midpoint of AB. Then angle $DMB = \dfrac{\pi}{2}$ radians as $BCDM$ is a rectangle.
Using trigonometry, $DM = 5 \sin x$ and $AM = 5 \cos x$
$\Rightarrow BC = DM = 5 \sin x$ and $AM = MB = CD = 5 \cos x$
Perimeter of trapezium, $P = AB + BC + CD + DA$
$= (2 \times 5 \cos x) + 5 \sin x + 5 \cos x + 5$
$= 15 \cos x + 5 \sin x + 5$
Let $15 \cos x + 5 \sin x = R \sin (x + \alpha)$
$= R(\sin x \cos \alpha + \cos x \sin \alpha)$
So $R \cos \alpha = 5$ and $R \sin \alpha = 15$
$\Rightarrow \tan \alpha = \dfrac{15}{5} = 3 \Rightarrow \alpha = \tan^{-1} 3 = 1.249...$
$R = \sqrt{15^2 + 5^2} = \sqrt{250} = 5\sqrt{10}$
So $15 \cos x + 5 \sin x = 5\sqrt{10} \sin (x + 1.249...)$
$\Rightarrow P = 5\sqrt{10} \sin (x + 1.249...) + 5$
So $R = 5\sqrt{10}$ and $\alpha = 1.249...$
[7 marks available — 1 mark for finding an expression for DM, 1 mark for finding an expression for AM, 1 mark for perimeter of trapezium in terms of x, 1 mark for expanding R sin (x + α), 1 mark for the correct value of α, 1 mark for finding the value of R, 1 mark for a correct substitution into the expression for P]

b) The perimeter is 17, so $17 = 5\sqrt{10}\sin(x + 1.249...) + 5$
$\Rightarrow 5\sqrt{10}\sin(x + 1.249...) = 12$
$\Rightarrow \sin(x + 1.249...) = \dfrac{12}{5\sqrt{10}}$ *[1 mark]*

As x is acute, $0 \leq x \leq \dfrac{\pi}{2}$. So you need to find solutions for
$1.249... \leq x + 1.249... \leq 2.819..$:

$x + 1.249... = \sin^{-1}\left(\dfrac{12}{5\sqrt{10}}\right)$

$\Rightarrow x + 1.249... = 0.8616...$ (outside range),
$x + 1.249... = \pi - 0.8616... = 2.279...$ *[1 mark]*
$\Rightarrow x = 2.279... - 1.249... = 1.030... = 1.03$ radians (3 s.f.)
[1 mark]

11 a) If the rate of change is proportional to the population,
then $\dfrac{dP}{dt} \propto P \Rightarrow \dfrac{dP}{dt} = kP$ for some constant k.
Separate the variables and integrate:
$\int \dfrac{1}{P}\,dP = \int k\,dt \Rightarrow \ln P = kt + \ln Q \Rightarrow P = e^{kt + \ln Q}$
$\Rightarrow P = e^{kt}e^{\ln Q} \Rightarrow P = Qe^{kt}$

[3 marks available — 1 mark for setting up a proportionality statement and converting it into an equation, 1 mark for separating variables and integrating both sides, 1 mark for rearranging to give function in the required form]

b) $Q = 5300$ *[1 mark]*
$876 = 5300e^{6k}$ *[1 mark]*
$\Rightarrow \dfrac{876}{5300} = e^{6k} \Rightarrow \ln\left|\dfrac{876}{5300}\right| = 6k$
$\Rightarrow 6k = -1.8000... \Rightarrow k = -0.3$ (1 d.p.) *[1 mark]*

c) The value of k is negative so the population is shrinking *[1 mark]*.

12 a) Use the distance between two points A and B:
$(2 - 9)^2 + (4 - 8)^2 + (1 - z)^2$ *[1 mark]* $= 5^2$ or 9^2
However, as $(2 - 9)^2 + (4 - 8)^2 > 25$ and $(1 - z)^2 \geq 0$, it cannot
be 5^2, so $(2 - 9)^2 + (4 - 8)^2 + (1 - z)^2 = 9^2$ *[1 mark]*
$49 + 16 + (1 - z)^2 = 81 \Rightarrow (1 - z)^2 = 16$
$\Rightarrow 1 - z = \pm 4$
so $z = 5$ *[1 mark]* or $z = -3$ *[1 mark]*

b) $|\overrightarrow{BC}| = 5$, since this is the other side length of the
parallelogram. So $\overrightarrow{BC} = 5(0.8\mathbf{j} + 0.6\mathbf{k})$ *[1 mark]*
$= 4\mathbf{j} + 3\mathbf{k}$ *[1 mark]*
$\overrightarrow{OC} = \overrightarrow{OB} + \overrightarrow{BC} = (9\mathbf{i} + 8\mathbf{j} + 5\mathbf{k}) + (4\mathbf{j} + 3\mathbf{k})$
So C has coordinates $(9, 8 + 4, 5 + 3) = (9, 12, 8)$ *[1 mark]*
You have to use z = 5 from part a).

c) $\overrightarrow{AC} = \overrightarrow{OC} - \overrightarrow{OA} = (9\mathbf{i} + 12\mathbf{j} + 8\mathbf{k}) - (2\mathbf{i} + 4\mathbf{j} + \mathbf{k})$ *[1 mark]*
$|\overrightarrow{AC}| = \sqrt{(9 - 2)^2 + (12 - 4)^2 + (8 - 1)^2}$ *[1 mark]*
$= \sqrt{7^2 + 8^2 + 7^2}$
$= \sqrt{162} = 9\sqrt{2}$ *[1 mark]*

13 a) Use integration by parts:
$u = \ln x, \dfrac{dv}{dx} = x^{-5}$ *[1 mark for both]*,
$\dfrac{du}{dx} = \dfrac{1}{x}, v = \dfrac{1}{-4}x^{-4}$ *[1 mark for both]*
$\int \dfrac{\ln x}{x^5}\,dx = \ln x \times \dfrac{1}{-4}x^{-4} - \int \dfrac{1}{-4}x^{-4} \times \dfrac{1}{x}\,dx$ *[1 mark]*
$= \dfrac{-1}{4x^4}\ln x + \dfrac{1}{4}\int x^{-5}\,dx = \dfrac{-1}{4x^4}\ln x - \dfrac{1}{16x^4} + c$ *[1 mark]*

b) $u = \sqrt{x - 1} \Rightarrow \dfrac{du}{dx} = \dfrac{1}{2\sqrt{x - 1}}$
$\Rightarrow 2\sqrt{x - 1}\,du = dx \Rightarrow dx = 2u\,du$ *[1 mark]*
Change the limits: $x = 10 \Rightarrow u = \sqrt{10 - 1} = \sqrt{9} = 3$
$x = 5 \Rightarrow u = \sqrt{5 - 1} = \sqrt{4} = 2$ *[1 mark for both limits]*
$u^2 = x - 1 \Rightarrow x = u^2 + 1$
$\int_5^{10} \dfrac{2x}{\sqrt{x - 1}}\,dx = \int_2^3 \dfrac{2(u^2 + 1)}{u} \times 2u\,du$ *[1 mark]*
$= \int_2^3 (4u^2 + 4)\,du = \left[\dfrac{4u^3}{3} + 4u\right]_2^3$ *[1 mark]*
$= \left(\dfrac{4}{3}(3^3) + 4(3)\right) - \left(\dfrac{4}{3}2^3 + 4(2)\right)$
$= 48 - \dfrac{56}{3} = \dfrac{88}{3}$ *[1 mark]*

14 a) Volume $= x^3$, so the price of concrete is $65x^3$ *[1 mark]*
Surface area $= 6x^2$, so the price of gold leaf is $900x^2$ *[1 mark]*
Total cost of materials is £250, so $65x^3 + 900x^2 = 250$
[1 mark]
So $900x^2 = 250 - 65x^3 \Rightarrow x^2 = \dfrac{250 - 65x^3}{900}$
$\Rightarrow x = \sqrt{\dfrac{250 - 65x^3}{900}}$
[1 mark for correct rearrangement]

b) Let $f(x) = x - \sqrt{\dfrac{250 - 65x^3}{900}}$
If x, the length of the cube, is between 0.5 and 1, there will be
a root of $f(x) = 0$ in the interval $0.5 < x < 1$, and a change of
sign for $f(x)$ between 0.5 and 1.

$f(0.5) = 0.5 - \sqrt{\dfrac{250 - 65(0.5)^3}{900}} = -0.018...$

$f(1) = 1 - \sqrt{\dfrac{250 - 65(1)^3}{900}} = 0.546...$ *[1 mark for both]*

There is a change of sign and the function is continuous
over this interval, so there is root in the interval $0.5 < x < 1$
[1 mark].

c) $x_2 = \sqrt{\dfrac{250 - 65(0.75)^3}{900}} = 0.49730... = 0.4973$ (4 d.p.)
[1 mark]
$x_3 = 0.51855...$
$x_4 = 0.51740...$
$x_5 = 0.51746... = 0.5175$ (4 d.p.) *[1 mark]*

d) Using x_5, concrete costs $65 \times (0.51746...)^3 = £9.0067...$
[1 mark]
Cost of cube $= £250$
$\dfrac{9.0067...}{250} = 0.0360...$
So concrete is 3.6% (1 d.p.) of the total cost *[1 mark]*.

e)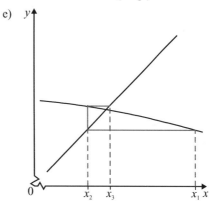

[1 mark for positions of x_2 and x_3 correct]
This is a convergent cobweb diagram.

Practice Exam Paper 2: Pure Mathematics and Statistics

1 a) $\dfrac{x^3 - 9x^2 + 14x}{x^2 - 4} \times \dfrac{x + 2}{x} = \dfrac{x^3 - 9x^2 + 14x}{(x + 2)(x - 2)} \times \dfrac{x + 2}{x}$
$= \dfrac{x^3 - 9x^2 + 14x}{x(x - 2)} = \dfrac{x^2 - 9x + 14}{(x - 2)} = \dfrac{(x - 2)(x - 7)}{(x - 2)}$
$= x - 7$
[3 marks available — 1 mark for factorising the numerator, 1 mark for factorising the denominator, 1 mark for fully simplified expression]

b) The points of intersection occur when $|x - 7| = -\dfrac{1}{2}x + 5$.
When $x > 7$:
$x - 7 = -\dfrac{1}{2}x + 5$ *[1 mark]* $\Rightarrow \dfrac{3}{2}x = 12 \Rightarrow x = 8$
When $x = 8$, $y = 8 - 7 = 1$
When $x < 7$:
$7 - x = -\dfrac{1}{2}x + 5$ *[1 mark]* $\Rightarrow \dfrac{1}{2}x = 2 \Rightarrow x = 4$
When $x = 4$, $y = 7 - 4 = 3$
So the lines intersect at $(4, 3)$ *[1 mark]* and $(8, 1)$ *[1 mark]*.

2 Rearranging the equation for x to make t the subject gives:
$t = x + 3$ *[1 mark]*
Substituting into the equation for y gives the Cartesian equation:
$y = (x + 3 + k)^2$ *[1 mark]*
Substitute in the coordinates $(-7, 0)$ to find the value of k:
$0 = (-7 + 3 + k)^2 \Rightarrow 3 + k = 7 \Rightarrow k = 4$ *[1 mark]*

3 a) The dog's range is a sector, bounded by the edge of the pen, the limit and the arc that is formed by the limit of its rope. The angle of the sector is the exterior angle of the regular hexagon: $360° \div 60° = \dfrac{\pi}{3}$ radians *[1 mark]*
So the area is $\dfrac{1}{2} \times 2^2 \times \dfrac{\pi}{3}$ *[1 mark]* $= \dfrac{2\pi}{3}$ m^2 *[1 mark]*

b) The hexagon is made up of 6 equilateral triangles of side length 2 m. Area of one of these triangles:
$\dfrac{1}{2} \times 2 \times 2 \times \sin\left(\dfrac{\pi}{3}\right)$ *[1 mark]* $= 2 \times \dfrac{\sqrt{3}}{2} = \sqrt{3}$ m^2
So the hexagon's area is $6\sqrt{3}$ m^2 *[1 mark]*
This means the cat can go in an area of
$(20 \times 8) - \dfrac{2\pi}{3} - 6\sqrt{3}$ *[1 mark]* $= 148$ m^2 (3 s.f.) *[1 mark]*

4 a) $\dfrac{dH}{dt} = -2\left(\dfrac{4\pi}{25}\right)\sin\left(\dfrac{4\pi t}{25}\right) = -\dfrac{8\pi}{25}\sin\left(\dfrac{4\pi t}{25}\right)$ *[1 mark]*

b) $\dfrac{dH}{dt}$ is the rate of change of the height of the buoy.
So when $t = 7$, $\dfrac{dH}{dt} = -\dfrac{8\pi}{25}\sin\left(\dfrac{4\pi \times 7}{25}\right)$ *[1 mark]*
$= 0.37007... = 0.370$ m/h (3 s.f.) *[1 mark]*

c) As $-1 \le \cos\left(\dfrac{4\pi t}{25}\right) \le 1$, the minimum value that $\cos\left(\dfrac{4\pi t}{25}\right)$ can take is -1 *[1 mark]*, and this occurs in the interval $0 < t < 12$. So the minimum height is $2(-1) + 4 = 2$ m *[1 mark]*.

5 Use the identity $\cot^2 x + 1 \equiv \operatorname{cosec}^2 x$
and substitute it into the equation:
$2(\cot^2 x + 1) + 5\cot x = 9 \Rightarrow 2\cot^2 x + 5\cot x - 7 = 0$ *[1 mark]*
Factorising gives $(2\cot x + 7)(\cot x - 1) = 0$ *[1 mark]*
$\cot x = -\dfrac{7}{2}$ or $\cot x = 1$ *[1 mark]*
$\Rightarrow \tan x = -\dfrac{2}{7}$ or $\tan x = 1$ *[1 mark]*
$x = -15.9°$ (1 d.p.) *[1 mark]* and $x = 45°$ *[1 mark]*.

6 a) Use the chain rule on the e^{x^2} term:
If $y = e^{x^2}$, then let $u = x^2$, so that $y = e^u$.
Then $\dfrac{du}{dx} = 2x$ and $\dfrac{dy}{du} = e^u$, which means
$\dfrac{dy}{dx} = \dfrac{dy}{du} \times \dfrac{du}{dx} = e^u \times 2x = 2xe^{x^2}$
So for $f(x) = 4x^3 + e^{x^2}$, $f'(x) = 12x^2 + 2xe^{x^2}$
[3 marks available — 1 mark for using the chain rule to differentiate e^{x^2}, 1 mark for each correct term in final answer]

b) Differentiate again to find the second derivative:
$\dfrac{d}{dx}12x^2 = 24x$
Use the product rule to differentiate $2xe^{x^2}$:
Let $u = 2x$, then $\dfrac{du}{dx} = 2$
Let $v = e^{x^2}$, then $\dfrac{dv}{dx} = 2xe^{x^2}$ (from part a)).
$\dfrac{dy}{dx} = u\dfrac{dv}{dx} + v\dfrac{du}{dx} = 2x(2xe^{x^2}) + e^{x^2}(2) = 4x^2e^{x^2} + 2e^{x^2}$
[1 mark for using the product rule correctly]
So $f''(x) = 24x + 4x^2e^{x^2} + 2e^{x^2}$
[1 mark for correct second derivative]
The curve is convex when $f''(x) > 0$,
i.e. when $24x + 4x^2e^{x^2} + 2e^{x^2} > 0$ *[1 mark]*
$e^{x^2} > 0$ and $4x^2 \ge 0$ for all x *[1 mark]*. If $x > 0$, $24x > 0$.
So for $x > 0$, $f''(x) > 0$, so the curve is convex when $x > 0$
[1 mark].

7 a)

x	0	0.5	1	1.5
y (3 d.p.)	1	**0.627**	0.415	0.270
y (5 d.p.)	1	**0.62671**	0.41497	0.26953

[1 mark for both values correct]

b) $\displaystyle\int_0^{1.5} \dfrac{e^{\sin x} + x}{(x+1)^3}\, dx \approx \dfrac{0.5}{2}(1 + 2(0.627 + 0.415) + 0.270)$
$= 0.8385$
[3 marks available — 1 mark for correct value of h (0.5), 1 mark for using the trapezium rule correctly, 1 mark for correct answer]

c) E.g. Sophie is correct that rounding will have contributed to the error, but as the graph is convex, this will have also contributed to the error *[1 mark for any comment about the graph being convex]*.

8 a) $\dfrac{dT}{dt}$ is the rate of heat loss with respect to time.
The difference in temperature is $(T - 15)$ °C.
$\dfrac{dT}{dt} \propto (T - 15)$, so $\dfrac{dT}{dt} = -k(T - 15)$, with the negative sign indicating that the temperature is decreasing *[1 mark]*.

b) Rearrange and integrate both sides of the equation from a)
$\displaystyle\int \dfrac{1}{T-15}\, dT = \int -k\, dt$
$\ln|T - 15| = -kt + C$ *[1 mark]*
$T - 15 = e^{-kt + C}$ *[1 mark]* $= Ae^{-kt}$ (where $A = e^C$)
Put in the initial condition $t = 0$, $T = 95$ to find A:
$95 - 15 = Ae^0 \Rightarrow A = 80$
$T = 80e^{-kt} + 15$ *[1 mark]*

c) Put in the given condition $t = 10$, $T = 55$ to find k:
$55 = 80e^{-10k} + 15$ *[1 mark]* $\Rightarrow e^{-10k} = 0.5$
$\Rightarrow -10k = \ln 0.5 \Rightarrow k = -\dfrac{\ln 0.5}{10}$ *[1 mark]*
$T = 80e^{\frac{\ln 0.5}{10} \times t} + 15$ *[1 mark]* $= 80(e^{\ln 0.5})^{\frac{t}{10}} + 15$
$\Rightarrow T = 80(0.5)^{\frac{t}{10}} + 15$ *[1 mark]*

d) E.g. the model assumes that the temperature in the kitchen remains constant *[1 mark for any sensible assumption]*.

9 a) Mean $= \dfrac{\sum fx}{\sum f} = \dfrac{21\,064\,931}{513\,242}$
$= 41.042... = 41.0$ years (3 s.f.) *[1 mark]*
Variance $= \dfrac{\sum fx^2}{\sum f} - \overline{x}^2 = \dfrac{1\,141\,982\,145}{513\,242} - 41.042...^2$
$= 540.518...$
So standard deviation $= \sqrt{540.518...} = 23.249...$
$= 23.2$ years (3 s.f.)
[1 mark for correct method, 1 mark for the correct answer]

b) Outliers are values that lie more than two standard deviations away from the mean.
So the lower limit is:
$41.042... - (2 \times 23.249...) = -5.455...$ years (impossible)
And the upper limit is:
$41.042... + (2 \times 23.249...) = 87.540... = 87.5$ years (3 s.f.)
[1 mark for both limits]
The lower limit is impossible, so there can only be outliers at the upper end of the data *[1 mark for any sensible comment]*.

c) E.g. There may have been a resident aged 88 or over in 2011 (an outlier) who has since died or moved away — this would mean that Elena's claim is wrong.
[1 mark for any sensible comment]

d) E.g. The mean for Newcastle upon Tyne is lower than the mean for County Durham, so people in Newcastle upon Tyne are generally younger. The standard deviation for Newcastle upon Tyne is smaller than the standard deviation for County Durham, so the ages in Newcastle upon Tyne are less spread out.
[2 marks available — 1 mark for a sensible comment comparing the means, 1 mark for a sensible comment comparing the standard deviations]

10 a) She has assumed that the probability of queuing for longer than 2 minutes is constant *[1 mark]*. She has assumed that her visits are independent *[1 mark]*.

Answers

b) Let X represent the number of days in her sample where Chloe has to queue for longer than 2 minutes. $X \sim B(50, 0.4)$, so $Y \sim N(50 \times 0.4, 50 \times 0.4 \times 0.6)$ $\Rightarrow Y \sim N(20, 12)$ *[1 mark]*
A normal approximation is appropriate because $n = 50$, which is quite large and $p = 0.4$, which is quite close to 0.5 *[1 mark]*
You could also have worked out np = 20 and n(p − 1) = 30 — as they're both greater than 5, a normal approximation is appropriate.

c) Systematic sampling *[1 mark]*

11 a) Moderate positive correlation — Local Authorities that have a higher percentage of the population aged 18-29 tend to have a higher percentage of the workforce that walk to work *[1 mark]*.

b) Region A = West Midlands, Region B = London
E.g. In Region B, there are clusters of data points with similar values — Local Authorities in London are more likely to have similar characteristics, as they are all city/urban areas.
[2 marks available — 1 mark for assigning the areas correctly, 1 mark for a sensible justification]
If you'd got the regions the wrong way round, but come up with a sensible explanation why, you'd probably still get the mark.

c) E.g. Yes, I think this was a sensible decision. This value is clearly an outlier, and doesn't reflect the percentages for the other Local Authorities in that region *[1 mark for a sensible comment]*.

12 a) E.g. From the shape of the histogram, it is clear that the data is not symmetrically distributed (there are more people at the younger end of the distribution), so a normal model would not be appropriate *[1 mark for a sensible comment]*.

b) If the data, A years, can be modelled using the given normal distribution, you would expect approximately 25% of the distribution to be less than $Q_1 = 45$ and 25% of the distribution to be greater than $Q_3 = 62$.
Using $A \sim N(54, 12.5^2)$, $P(A < 45) = 0.2358$ (4 d.p.) and $P(A > 62) = 0.2611$ (4 d.p.).
Both of these probabilities are close to 0.25, so this normal distribution seems like it could be a suitable model.
[2 marks available — 1 mark for finding P(A < 45) and P(A > 62), 1 mark for a sensible conclusion]
You could have done this one slightly differently — e.g. by finding Q_1 and Q_3 for the given normal distribution and comparing them to the values Leanne calculated.

13 a) $P(J < 493) = 0.05$ and $P(J > 502) = 0.025$ *[1 mark]*

$P(J < 493) = 0.05 \Rightarrow P\left(Z < \dfrac{493 - \mu}{\sigma}\right) = 0.05$
$\Rightarrow \dfrac{493 - \mu}{\sigma} = -1.6449 \Rightarrow \mu - 493 = 1.6449\sigma$ *[1 mark]*
$P(J > 502) = 0.025 \Rightarrow P(J < 502) = 0.975$
$\Rightarrow P\left(Z < \dfrac{502 - \mu}{\sigma}\right) = 0.975$
$\Rightarrow \dfrac{502 - \mu}{\sigma} = 1.9600 \Rightarrow 502 - \mu = 1.9600\sigma$ *[1 mark]*
Solve these simultaneously:
$(\mu - 493) + (502 - \mu) = 1.6449\sigma + 1.9600\sigma$
$\Rightarrow 9 = 3.6049\sigma \Rightarrow \sigma = 2.4966... = 2.50$ (3 s.f.) *[1 mark]*
$\mu = 493 + 1.6449 \times 2.4966... = 497.10...$
$= 497$ (3 s.f.) *[1 mark]*

b) From part a), $J \sim N(497, 2.50^2)$ *[1 mark]*.
The probability that a carton fails to meet the manufacturer's standards is $P(J < 492)$. From your calculator, this is $0.02275... = 0.0228$ (4 d.p.) *[1 mark]*

c) $H_0: \mu = 497$ *[1 mark]*, $H_1: \mu \neq 497$ *[1 mark]*
— so this is a 2-tail test.
Under H_0, $J \sim N(497, 2.50^2)$, so the sample mean,
$\bar{J} \sim N\left(497, \dfrac{2.50^2}{20}\right)$ *[1 mark]*, and $Z = \dfrac{\bar{J} - 497}{\frac{2.50}{\sqrt{20}}} \sim N(0, 1)$.
$\bar{j} = 498.7$, so $z = \dfrac{498.7 - 497}{\frac{2.50}{\sqrt{20}}} = 3.0410...$ *[1 mark]*

This is a 2-tail test, and as $z > 0$, you're interested in the upper tail — so you need to find z such that $P(Z < z) = 0.005$.
$P(Z > 2.5758) = 0.005$, so the critical value is 2.5758 and the critical region is $Z > 2.5758$. Since $3.0410... > 2.5758$ *[1 mark]*, the result is significant *[1 mark]*, so there is evidence at the 1% level of significance to reject H_0 in favour of the alternative hypothesis that the mean volume of juice in a carton has changed *[1 mark]*.
You could have found P(Z > 3.0410...) instead — you get a value of 0.00117..., which is less than 0.005 so is significant. You'd get the marks for either method — use whichever one you prefer.

14 a) Using the conditional probability formula:
$P(R|S) = \dfrac{P(R \cap S)}{P(S)} = \dfrac{4}{9} \Rightarrow P(R \cap S) = \dfrac{4}{9}P(S)$ *[1 mark]*
$P(S|R) = \dfrac{P(R \cap S)}{P(R)} = \dfrac{4}{11} \Rightarrow P(R \cap S) = \dfrac{4}{11}P(R)$ *[1 mark]*
So $\dfrac{4}{9}P(S) = \dfrac{4}{11}P(R) \Rightarrow P(S) = \dfrac{9}{11}P(R)$ *[1 mark]*
Now, using the addition law:
$P(R \cup S) = P(R) + P(S) - P(R \cap S)$
$\dfrac{8}{10} = P(R) + \dfrac{9}{11}P(R) - \dfrac{4}{11}P(R)$ *[1 mark]*
$\dfrac{8}{10} = \dfrac{16}{11}P(R) \Rightarrow P(R) = \dfrac{11}{20} = 0.55$
And $P(S) = \dfrac{9}{11}P(R) = \dfrac{9}{11} \times \dfrac{11}{20} = \dfrac{9}{20} = 0.45$
[1 mark for both P(R) and P(S)]

b) $P(R \cap S) = \dfrac{4}{9}P(S) = \dfrac{4}{9} \times \dfrac{9}{20} = \dfrac{1}{5} = 0.2$
$P(R) \times P(S) = \dfrac{11}{20} \times \dfrac{9}{20} = \dfrac{99}{400} = 0.2475$
[1 mark for both probabilities]
So $P(R \cap S) \neq P(R) \times P(S)$, meaning that the events R and S are not independent *[1 mark]*.
There are other ways to check whether or not these are independent events — for example, you could show that $P(R) \neq P(R|S)$.

c) $P(S \cap T) = P(S) \times P(T) = 0.45 \times 0.15 = 0.0675$
Now work everything out using the known probabilities:

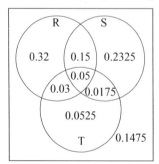

[4 marks for a completely correct Venn diagram, otherwise 1 mark for a correct method to find P(S ∩ T), 1 mark for P(S ∩ T) = 0.0675, 1 mark for P(R ∩ S ∩ T') or P(R ∩ T ∩ S') correct]

15 a) Alice's result shows moderately strong positive correlation, which suggests that the number of bacteria tends to be higher in deeper ponds *[1 mark]*.
Ben's result is greater than 1, which is not a valid answer, so he must have made an error in his calculation *[1 mark]*.

b) While the results suggest that they are linked, they do not mean that deeper ponds cause higher numbers of bacteria. There could be a third variable linking them, or it could just be coincidence. *[1 mark]*

c) $H_0: \rho = 0$, $H_1: \rho > 0$ *[1 mark]*.
The test statistic is $r = 0.71$. The critical value is 0.6215. Since $0.71 > 0.6215$, the result is significant.
There is evidence at the 5% significance level to reject H_0 and to support the alternative hypothesis that the depth of a pond and the number of this type of bacteria in the pond are positively correlated *[1 mark]*.

Practice Exam Paper 3: Pure Mathematics and Mechanics

1 Start with the LHS:

$\sec 2\theta \equiv \dfrac{1}{\cos 2\theta}$

Use the double angle formula $\cos 2\theta \equiv \cos^2 \theta - \sin^2 \theta$:

$\equiv \dfrac{1}{\cos^2 \theta - \sin^2 \theta}$ *[1 mark]*

Divide the numerator and denominator by $\cos^2 \theta$:

$\equiv \dfrac{\sec^2 \theta}{1 - \tan^2 \theta}$ *[1 mark]*

Use the trig identity $\tan^2 \theta \equiv \sec^2 \theta - 1$:

$\equiv \dfrac{\sec^2 \theta}{1 - (\sec^2 \theta - 1)}$

$\equiv \dfrac{\sec^2 \theta}{2 - \sec^2 \theta}$ as required *[1 mark for correct rearrangement]*

2 a)

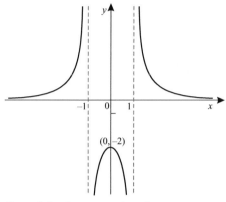

[1 mark for the correct shape]

Asymptotes at $x = 1$, $x = -1$ *[1 mark]* and $y = 0$ *[1 mark]*, y-intercept at $(0, -2)$ *[1 mark]*

b)

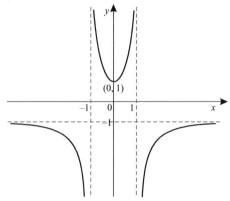

[1 mark for correction transformations (a reflection in the x-axis and translation of −1 in the y-direction)]

Asymptotes at $x = 1$, $x = -1$ and $y = -1$ *[1 mark for all three]*, y-intercept at $(0, 1)$ *[1 mark]*

3 a) $\displaystyle\int_0^{2\pi} 2\cos\dfrac{x}{4}\, dx$ is the area under the curve between $x = 0$ and $x = 2\pi$. The area of a rectangle with vertices $(0, 0)$ $(0, 2)$ $(2\pi, 0)$ and $(2\pi, 2)$ would have area $2\pi \times 2 = 4\pi$.

The area beneath the curve and between $x = 0$ and $x = 2\pi$ fits inside this rectangle and has an area less than the rectangle, so Lily's answer must be too big *[1 mark]*.

b) $\displaystyle\int_0^{2\pi} 2\cos\dfrac{x}{4}\, dx = \left[8\sin\dfrac{x}{4}\right]_0^{2\pi} = \left[8\sin\dfrac{\pi}{2}\right] - [8\sin 0] = 8$

[3 marks available — 1 mark for integrating correctly, 1 mark for correct handling of the limits, 1 mark for correct answer]

4 a) The number of minutes of practice on the nth day is given by

$u_n = a + (n-1)d = 60 + (n-1)10 = 50 + 10n$ *[1 mark]*

4 hours 40 minutes is 280 minutes, so

$50 + 10n = 280$ *[1 mark]* $\Rightarrow n = 23$, i.e. day 23 *[1 mark]*

b) The number of minutes of practice on the nth day is given by

$u_n = ar^{n-1} = 100 \times 1.04^{n-1}$ *[1 mark]*

$100 \times 1.04^{n-1} > 280 \Rightarrow 1.04^{n-1} > 2.8$ *[1 mark]*

Taking logs gives: $\log 1.04^{n-1} > \log 2.8$

$\Rightarrow (n-1)\log 1.04 > \log 2.8$ *[1 mark]*

$\Rightarrow n - 1 > \dfrac{\log 2.8}{\log 1.04} \Rightarrow n > 27.3$ (3 s.f.)

So day 28 *[1 mark]*

c) Pianist: $\dfrac{1}{2} \times 30[(2 \times 60) + (29 \times 10)] = 6150$ minutes *[1 mark]*

Violinist: $\dfrac{100(1.04^{30} - 1)}{1.04 - 1} = 5608.5$ minutes *[1 mark]*

The pianist will practise for longer over the 30 days *[1 mark]*.

5 $\dfrac{d}{dx}(5x^3 + 2x^2 + 6) = 15x^2 + 4x$, and $45x^2 + 12x = 3(15x^2 + 4x)$

So $\displaystyle\int \dfrac{45x^2 + 12x}{5x^3 + 2x^2 + 6}\, dx = \int \dfrac{3f'(x)}{f(x)}\, dx = 3\ln|f(x)| + C$

$= 3\ln|5x^3 + 2x^2 + 6| + C$

[3 marks available — 1 mark for recognising that the numerator is 3 × the derivative of the denominator, 1 mark for integrating using the correct rule, 1 mark for the correct answer, including the constant]

6 a) $t = 1 \Rightarrow x = 1^3 + 2 = 3$ and $y = 1^2 + 2 = 3$, so the coordinates of P are $(3, 3)$ *[1 mark]*.

b) $\dfrac{dx}{dt} = 3t^2$, $\dfrac{dy}{dt} = 2t$ *[1 mark for both]*

Using the chain rule: $\dfrac{dy}{dx} = \dfrac{dy}{dt} \div \dfrac{dx}{dt} = \dfrac{2t}{3t^2} = \dfrac{2}{3t}$ *[1 mark]*.

At P, $t = 1$, so $\dfrac{dy}{dx} = \dfrac{2}{3}$ *[1 mark]*.

Using the equation of a straight line formula at P $(3, 3)$:

$y - 3 = \dfrac{2}{3}(x - 3) \Rightarrow y = \dfrac{2}{3}x + 1$ *[1 mark]*.

7 Use the quotient rule: $u = 3 + \sin x \Rightarrow \dfrac{du}{dx} = \cos x$ *[1 mark]*

$v = (2x + 1)^4 \Rightarrow \dfrac{dv}{dx} = 4(2)(2x + 1)^3 = 8(2x + 1)^3$ *[1 mark]*

$f'(x) = \dfrac{(2x + 1)^4 \cos x - 8(3 + \sin x)(2x + 1)^3}{(2x + 1)^8}$

$= \dfrac{(2x + 1)\cos x - 8(3 + \sin x)}{(2x + 1)^5}$

[1 mark for use of quotient rule, 1 mark for correct answer]

8 Differentiate the equation:

$2x - y - x\dfrac{dy}{dx} = 6y^2\dfrac{dy}{dx} \Rightarrow \dfrac{dy}{dx} = \dfrac{2x - y}{x + 6y^2}$

To find the gradient of the tangent at $(2, 1)$,

substitute $x = 2$ and $y = 1$ into the expression for $\dfrac{dy}{dx}$:

$\dfrac{dy}{dx} = \dfrac{2(2) - 1}{2 + 6(1)^2} = \dfrac{3}{8}$.

So the gradient of the normal to the curve at $(2, 1)$ is $-1 \div \dfrac{3}{8} = -\dfrac{8}{3}$

Now find the equation of the normal using $y - y_1 = m(x - x_1)$:

$y - 1 = -\dfrac{8}{3}(x - 2) \Rightarrow y = -\dfrac{8}{3}x + \dfrac{19}{3}$

When this line crosses the x-axis, $y = 0$, so $0 = -\dfrac{8}{3}x + \dfrac{19}{3} \Rightarrow$

$x = \dfrac{19}{8}$, so the coordinates of the point of intersection are $\left(\dfrac{19}{8}, 0\right)$.

[7 marks available — 1 mark for differentiating all terms correctly, 1 mark for the correct expression for dy/dx, 1 mark for substituting x = 2 and y = 1, 1 mark for the gradient of the tangent, 1 mark for the gradient of the normal, 1 mark for the equation of the normal, 1 mark for the correct coordinates]

9 a) Set the equations equal to each other and rearrange:

$\dfrac{1}{x^3} = x + 3 \Rightarrow 1 = x^4 + 3x^3$

$\Rightarrow x^4 + 3x^3 - 1 = 0$ *[1 mark]*

b) Let $f(x) = x^4 + 3x^3 - 1$, then $f'(x) = 4x^3 + 9x^2$ *[1 mark]*

Using the Newton-Raphson formula

$x_{n+1} = x_n - \dfrac{f(x_n)}{f'(x_n)} = x_n - \dfrac{x_n^4 + 3x_n^3 - 1}{4x_n^3 + 9x_n^2}$ *[1 mark]*

$x_1 = 0.6$

$x_2 = 0.6 - \dfrac{0.6^4 + 3(0.6^3) - 1}{4(0.6^3) + 9(0.6^2)}$

$= 0.65419... = 0.6542$ (4 d.p.)

$x_3 = 0.65419... - \dfrac{0.65419...^4 + 3(0.65419...^3) - 1}{4(0.65419...^3) + 9(0.65419...^2)}$

$= 0.64955... = 0.6496$ (4 d.p.) *[1 mark for both x_2 and x_3]*

c) If $x = 0.650$ to 3 d.p., then the upper and lower bounds of x are 0.6505 and 0.6495 respectively *[1 mark]*: f(0.6495) = −0.0000671.... and f(0.6505) = 0.004833... *[1 mark for both values]* f(0.6495) < 0 and f(0.6505) > 0, which means there's a change of sign in the interval *[1 mark for pointing out the change of sign]*, and f(x) is continuous, so $\alpha = 0.650$ to 3 d.p.

10 a) Resolving forces vertically: $R_P + R_Q = 6g + 4g$ $R_P = R_Q$, so: $2R = 98 \Rightarrow R = 49$ N *[1 mark]* Taking moments about A: clockwise = anticlockwise $(6g \times 3) + (4g \times x) = (49 \times 1) + (49 \times 4)$ *[1 mark]* $\Rightarrow 176.4 + 39.2x = 245 \Rightarrow x = 1.75$ *[1 mark]* *You could take moments about any point — as long as your working and answer are correct, you'll get all the marks.*

b) For the maximum value of M, the rod will be on the point of tipping about Q, so R_P will be 0. *[1 mark]* Taking moments about Q: clockwise = anticlockwise $(Mg \times 2) = (6g \times 1) + (4g \times 2.25)$ $\Rightarrow 2M = 6 + 9 = 15 \Rightarrow M = 7.5$ *[1 mark]*

11 a) $\mathbf{v} = \dot{\mathbf{r}} = (3t^2 − 12t + 4)\mathbf{i} + (7 − 8t)\mathbf{j}$ At $t = 5$: $\mathbf{v} = (3(5^2) − 12(5) + 4)\mathbf{i} + (7 − 8(5))\mathbf{j} = 19\mathbf{i} − 33\mathbf{j}$ as required. *[3 marks available — 1 mark for attempting to differentiate the position vector, 1 mark for a correct expression for velocity at time t, and 1 mark for substituting t = 5 into this expression to obtain the correct velocity]*

b) When P is moving south, component of velocity in direction of $\mathbf{i}$ will be zero, i.e. $3t^2 − 12t + 4 = 0$ *[1 mark]* Solve using the quadratic formula: $t = \dfrac{−(−12) \pm \sqrt{(−12)^2 − 4 \times 3 \times 4}}{2 \times 3}$ *[1 mark]* $= 0.36700...$ or $3.63299...$ *[1 mark]* When P is moving south, the component of velocity in the direction of $\mathbf{j}$ will be negative, so find the velocity of P at the two values of t above, and see which is negative: $\mathbf{v}(0.36700...) = (7 − 8(0.36700...))\mathbf{j} = 4.063...\mathbf{j}$ (so P is moving due north) $\mathbf{v}(3.63299...) = (7 − 8(3.63299...))\mathbf{j} = −22.06...\mathbf{j}$ So P is moving due south when $t = 3.63$ s (3 s.f.) *[1 mark]* *Also, when P is moving due south, 7 − 8t must be negative, so $7 − 8t < 0 \Rightarrow t > \dfrac{7}{8}$, so t cannot be 0.3670...*

c) To find force on P, will need to use $\mathbf{F} = m\mathbf{a}$, so first find $\mathbf{a}$: $\mathbf{a} = \dot{\mathbf{v}} = (6t − 12)\mathbf{i} − 8\mathbf{j}$ *[1 mark]* When $t = 3$, $\mathbf{a} = (18 − 12)\mathbf{i} − 8\mathbf{j} = 6\mathbf{i} − 8\mathbf{j}$ *[1 mark]* So, $\mathbf{F} = m\mathbf{a} = 2.5(6\mathbf{i} − 8\mathbf{j}) = 15\mathbf{i} − 20\mathbf{j}$ *[1 mark]* Magnitude of $\mathbf{F} = \sqrt{15^2 + (−20)^2}$ *[1 mark]* $= 25$ N *[1 mark]*

12 a) $\mathbf{u} = (40\mathbf{i} + 108\mathbf{j})$ kmh⁻¹, $\mathbf{a} = (4\mathbf{i} + 12\mathbf{j})$ kmh⁻², $t = \dfrac{1}{2}$ h, Use $\mathbf{s} = \mathbf{u}t + \dfrac{1}{2}\mathbf{a}t^2$: $\mathbf{s} = \dfrac{1}{2}(40\mathbf{i} + 108\mathbf{j}) + \left[\dfrac{1}{2} \times \left(\dfrac{1}{2}\right)^2 \times (4\mathbf{i} + 12\mathbf{j})\right]$ *[1 mark]* $\mathbf{s} = (20.5\mathbf{i} + 55.5\mathbf{j})$ km *[1 mark for 20.5i, 1 mark for 55.5j]* *Make sure you keep track of the units you're using — you have to convert the time from minutes to hours to match the units of the velocity and acceleration.*

b) $\mathbf{u} = (60\mathbf{i} + 120\mathbf{j})$, $t = \dfrac{1}{4}$, $\mathbf{s} = (20.5\mathbf{i} + 55.5\mathbf{j}) − (2.5\mathbf{i} + 25.5\mathbf{j}) = (18\mathbf{i} + 30\mathbf{j})$ *[1 mark]* Use $\mathbf{s} = \mathbf{u}t + \dfrac{1}{2}\mathbf{a}t^2$: $(18\mathbf{i} + 30\mathbf{j}) = \dfrac{1}{4}(60\mathbf{i} + 120\mathbf{j}) + \dfrac{1}{2} \times \left(\dfrac{1}{4}\right)^2 \times \mathbf{a}$ *[1 mark]* $\Rightarrow \dfrac{1}{32}\mathbf{a} = (18\mathbf{i} + 30\mathbf{j}) − (15\mathbf{i} + 30\mathbf{j})$ $\Rightarrow \mathbf{a} = 32(3\mathbf{i} + 0\mathbf{j}) = 96\mathbf{i}$ So the magnitude of the acceleration is 96 kmh⁻² *[1 mark]*.

c) $\mathbf{u} = (30\mathbf{i} − 40\mathbf{j})$, $\mathbf{a} = (5\mathbf{i} + 8\mathbf{j})$, $\mathbf{v} = \mathbf{v}$, $t = ?$, $\mathbf{s} = ?$, Use $\mathbf{v} = \mathbf{u} + \mathbf{a}t$ to find the time that G is moving parallel to $\mathbf{i}$ (the unit vector in the direction of east): $\mathbf{v} = (30\mathbf{i} − 40\mathbf{j}) + t(5\mathbf{i} + 8\mathbf{j})$ *[1 mark]*

When G is moving parallel to $\mathbf{i}$, the $\mathbf{j}$-component of its velocity is zero, so: $0 = −40 + 8t$ *[1 mark]* $\Rightarrow t = 5$ Check that the i-component is positive: $30 + 5(5) = 55$, so G is moving due east at $t = 5$ *[1 mark]*. Now use $\mathbf{s} = \dfrac{1}{2}(\mathbf{u} + \mathbf{v})t$ to find the position of G at this time: $\mathbf{s} = \dfrac{1}{2}(30\mathbf{i} − 40\mathbf{j} + (30\mathbf{i} − 40\mathbf{j}) + 5(5\mathbf{i} + 8\mathbf{j})) \times 5$ *[1 mark]* $\mathbf{s} = \dfrac{5}{2}(85\mathbf{i} − 40\mathbf{j}) = (212.5\mathbf{i} − 100\mathbf{j})$ km *[1 mark for 212.5i, 1 mark for −100j]*.

13 a) Consider vertical motion, taking up as positive: $a = −g$, $u = U \sin \alpha$ *[1 mark]*, $v = 0$, $s = ?$: Use $v^2 = u^2 + 2as$: $0 = U^2 \sin^2 \alpha − 2gs$ *[1 mark]* $\Rightarrow s = \dfrac{U^2 \sin^2 \alpha}{2g}$ *[1 mark]*. The golf ball is initially 0.5 m above the ground, so the maximum height, h, it reaches is: $h = \dfrac{1}{2} + \dfrac{U^2 \sin^2 \alpha}{2g} = \dfrac{g + U^2 \sin^2 \alpha}{2g}$ m, as required *[1 mark]*.

b) You first need to find the horizontal and vertical components of the golf ball's motion when it lands. Vertically, taking up as positive: $u = U \sin \alpha$, $a = −g$, $s = −0.5$, $v = v_V$. Use $v^2 = u^2 + 2as$: $v_V^2 = U^2\sin^2\alpha + g$ *[1 mark]*. Horizontally, $v_H = u_H = U \cos \alpha$, as acceleration is zero *[1 mark]*. So, $V = \sqrt{v_V^2 + v_H^2} = \sqrt{U^2 \sin^2 \alpha + g + U^2 \cos^2 \alpha}$ *[1 mark]* $= \sqrt{U^2(\sin^2\alpha + \cos^2\alpha) + g}$ *[1 mark]* $= \sqrt{U^2 + g}$ ms⁻¹, as required *[1 mark]*.

c) E.g. The effects of air resistance could be factored in. The beach ball should not be modelled as a particle — its diameter should be taken into account when finding distances. *[2 marks available — 1 mark for each sensible suggestion]*

14 a) $\theta = \tan^{-1}\left(\dfrac{3}{4}\right)$ so $\tan \theta = \dfrac{3}{4}$. This means that the opposite and adjacent sides of a right-angled triangle are 3 and 4, so the hypotenuse is $\sqrt{3^2 + 4^2} = 5$. This gives $\sin \theta = \dfrac{3}{5} = 0.6$ and $\cos \theta = \dfrac{4}{5} = 0.8$ Resolving forces parallel to the plane ($\nearrow$) for A, where T is the tension in the string: $F_{net} = ma \Rightarrow T − mg \sin \theta = ma$ $\Rightarrow ma = T − 5.88m$ ① *[1 mark]* Resolving forces perpendicular to the plane ($\nearrow$) for B: $R − 2mg \cos 70° = 0 \Rightarrow R = 19.6m \cos 70°$ *[1 mark]* Friction is limiting, so: $F = \mu R = 0.45 \times 19.6m \cos 70° = 8.82m \cos 70°$ *[1 mark]* Resolving forces parallel to the plane ($\searrow$) for B: $F_{net} = ma \Rightarrow 2mg \sin 70° − T − F = 2ma$ ② *[1 mark]* Now consider ① + ②: $3ma = T − 5.88m + 2mg \sin 70° − T − F$ $= 2mg \sin 70° − 5.88m − 8.82m \cos 70°$ *[1 mark]* Divide through by $3m$ to find a: $a = \dfrac{2g \sin 70° − 5.88 − 8.82 \cos 70°}{3}$ $= 3.1737... = 3.17$ ms⁻² (3 s.f.) *[1 mark]*

b) A is on the point of sliding down the plane, so both particles are in equilibrium. Resolving forces parallel to the plane ($\swarrow$) for A: $10g \sin \theta − T = 0 \Rightarrow T = 58.8$ N *[1 mark]* Resolving forces perpendicular to the plane ($\nearrow$) for B: $R − 2mg \cos 70° = 0 \Rightarrow R = 2mg \cos 70°$ Again, friction is limiting, so: $F = \mu R = 0.45 \times 2mg \cos 70°$ *[1 mark]* Resolving forces parallel to the plane ($\searrow$) for B (where the friction now acts down the plane): $T − F − 2mg \sin 70° = 0$ $\Rightarrow 58.8 − 0.45 \times 2mg \cos 70° − 2mg \sin 70° = 0$ *[1 mark]* $\Rightarrow m = \dfrac{58.8}{2(0.45g \cos 70° + g \sin 70°)} = 2.74$ kg (3 s.f.) *[1 mark]*

Answers

Formula Sheet

These formulas are the ones you'll be given in your exam.
Make sure you know exactly **when you need them** and **how to use them**.

Arithmetic Series

$$S_n = \frac{1}{2}n(a + l)$$

$$= \frac{1}{2}n[2a + (n - 1)d]$$

Geometric Series

$$S_n = \frac{a(1 - r^n)}{1 - r}$$

$$S_\infty = \frac{a}{1 - r} \text{ for } |r| < 1$$

Binomial Series

$$(a + b)^n = a^n + {}^nC_1 a^{n-1}b + {}^nC_2 a^{n-2}b^2 + \ldots + {}^nC_r a^{n-r}b^r + \ldots + b^n \quad (n \in \mathbb{N})$$

$$\text{where } {}^nC_r = \binom{n}{r} = \frac{n!}{r!(n - r)!}$$

$$(1 + x)^n = 1 + nx + \frac{n(n - 1)}{2!}x^2 + \ldots + \frac{n(n - 1)\ldots(n - r + 1)}{r!}x^r + \ldots \quad (|x| < 1, n \in \mathbb{R})$$

Differentiation

f(x)	f'(x)
$\tan kx$	$k \sec^2 kx$
$\sec x$	$\sec x \tan x$
$\cot x$	$-\text{cosec}^2 x$
$\text{cosec } x$	$-\text{cosec } x \cot x$

Quotient rule: for $y = \dfrac{u}{v}$, $\dfrac{dy}{dx} = \dfrac{v\dfrac{du}{dx} - u\dfrac{dv}{dx}}{v^2}$

Differentiation from first principles: $f'(x) = \lim\limits_{h \to 0} \dfrac{f(x + h) - f(x)}{h}$

Integration

$$\int \frac{f'(x)}{f(x)} \, dx = \ln|f(x)| + c$$

$$\int f'(x)(f(x))^n \, dx = \frac{1}{n + 1}(f(x))^{n+1} + c$$

Integration by parts: $\int u\dfrac{dv}{dx}dx = uv - \int v\dfrac{du}{dx}dx$

Formula Sheet

Trigonometry

Small angle approximations: $\sin \theta \approx \theta$, $\cos \theta \approx 1 - \frac{1}{2}\theta^2$, $\tan \theta \approx \theta$, where θ is measured in radians.

Trigonometric identities:

$$\sin (A \pm B) \equiv \sin A \cos B \pm \cos A \sin B$$

$$\cos (A \pm B) \equiv \cos A \cos B \mp \sin A \sin B$$

$$\tan (A \pm B) \equiv \frac{\tan A \pm \tan B}{1 \mp \tan A \tan B} \quad (A \pm B \neq (k + \tfrac{1}{2})\pi)$$

Numerical Methods

Trapezium rule: $\int_a^b y\, \mathrm{d}x \approx \frac{1}{2}h\{(y_0 + y_n) + 2(y_1 + y_2 + \ldots + y_{n-1})\}$, where $h = \dfrac{b-a}{n}$

The Newton-Raphson iteration for solving $f(x) = 0$: $x_{n+1} = x_n - \dfrac{f(x_n)}{f'(x_n)}$

Probability

$$P(A \cup B) = P(A) + P(B) - P(A \cap B)$$

$$P(A \cap B) = P(A)P(B|A) = P(B)P(A|B) \quad \text{or} \quad P(A|B) = \frac{P(A \cap B)}{P(B)}$$

Standard Deviation

$$\sqrt{\frac{\sum (x - \overline{x})^2}{n}} = \sqrt{\frac{\sum x^2}{n} - \overline{x}^2} \quad \text{or} \quad \sqrt{\frac{\sum f(x - \overline{x})^2}{\sum f}} = \sqrt{\frac{\sum fx^2}{\sum f} - \overline{x}^2}$$

The Binomial Distribution

If $X \sim B(n, p)$, then $P(X = x) = \dbinom{n}{x}p^x(1 - p)^{n-x}$

Mean of $X = np$ Variance of $X = np(1 - p)$

Formula Sheet

Hypothesis Testing for the Mean of a Normal Distribution

$$\text{If } X \sim N(\mu, \sigma^2), \text{ then } \overline{X} \sim N\!\left(\mu, \frac{\sigma^2}{n}\right) \text{ and } \frac{\overline{X} - \mu}{\sigma / \sqrt{n}} \sim N(0, 1)$$

Percentage Points of the Normal Distribution

If Z has a normal distribution with mean 0 and variance 1 then, for each value of p, the table gives the value of z such that $P(Z \le z) = p$.

p	0.75	0.90	0.95	0.975	0.99	0.995	0.9975	0.999	0.9995
z	0.674	1.282	1.645	1.960	2.326	2.575	2.807	3.090	3.291

Kinematics

Motion in a straight line:

$$v = u + at$$

$$s = ut + \frac{1}{2}at^2$$

$$s = \frac{1}{2}(u + v)t$$

$$v^2 = u^2 + 2as$$

$$s = vt - \frac{1}{2}at^2$$

Motion in two dimensions:

$$\mathbf{v} = \mathbf{u} + \mathbf{a}t$$

$$\mathbf{s} = \mathbf{u}t + \frac{1}{2}\mathbf{a}t^2$$

$$\mathbf{s} = \frac{1}{2}(\mathbf{u} + \mathbf{v})t$$

$$\mathbf{s} = \mathbf{v}t - \frac{1}{2}\mathbf{a}t^2$$

MRQ71